Modelling with Ordinary Differential Equations

Modelling with Ordinary Differential Equations

T.P. Dreyer

Department of Applied Mathematics
University of Stellenbosch
Stellenbosch, South Africa

CRC Press
Boca Raton Ann Arbor London Tokyo

Library of Congress Cataloging-in-Publication Data

Dreyer, T. P.
Modelling with ordinary differential equations / T. P. Dreyer.
p. cm.
Includes bibliographical references and index.
ISBN 0-8493-8636-5
1. Mathematical models. 2. Differential equations. I. Title.
QA401.D74 1993
511'.—dc20 93-9636
CIP

Direct all inquiries to CRC Press, Inc., 2000 Corporate Blvd., N.W., Boca Raton, Florida 33431.

International Standard Book Number 0-8493-8636-5

Library of Congress card number 93-9636

Printed in the United States 1 2 3 4 5 6 7 8 9 0
Printed on acid-free paper

– CONTENTS –

PREFACE

This is a book about mathematical modelling. The aim of the book is threefold:

1. To teach the basic skills of modelling
2. To introduce a basic tool for modelling, namely, ordinary differential equations
3. To show the wide scope of applications which can be modelled with ordinary differential equations

The book is not a reference manual and does not pretend to cover completely either one of the fields of modelling or ordinary differential equations. It is primarily meant as an elementary text to be used in a lecture room. It can, however, also be read as a self-study program since each model is developed from first principles, examples are provided in the text to clarify new definitions, and carefully chosen exercises are given in each section to strengthen the understanding of the material and the skill to construct models. Supplementary references are also given at the end of each section to stimulate further reading.

The only prerequisite for this book is a fair background in differential and integral calculus of one variable (including Taylor series) and the basic notions of a partial derivative and a vector in two or three dimensions.

The first version of this book in the form of classroom notes was written in 1972. As the material was tested in the classroom over two decades, some material was discarded, new models were added, and the selection of exercises and projects was refined. This book is the end result of a marriage between the academic goals set by the teacher and the feedback through questionnaires and examinations of the students. The favorable comments of ex-students working as applied mathematicians in industry indicate that the choice of material and the level of sophistication in the book are satisfactory.

Some mathematicians think that modelling is a rather obvious exercise, and hence a textbook should concentrate on mathematical theory and leave the rest to common sense. Anyone who has done some mathematical work for industry knows that this viewpoint is false and that the mathematical solution of some mathematical equations is only a part (many times the easiest part) of the modelling process. In this book, the theory of differential equations and the modelling process are interwoven purposefully, not only to convey the importance of both, but also to highlight the essential interaction between them.

There are a number of interesting models in the subsequent pages, and it is hoped that the reader shall enjoy reading about them, but to learn something about modelling the exercises must be done. It is nice to watch a good game of tennis, but as long as you are sitting in the stand, your game will not improve! You can check your answers at the back of the book.

Apart from the exercises, there is a section on projects at the end of each chapter. These are mainly intended for the computer-literate reader, and can also be used in tutorials.

The proofs of theorems appear in a section at the end of each chapter to facilitate the flow of the argument in the model-building process. It does not mean that the proofs are unimportant or that they should be left out.

The book consists of seven chapters. In the first chapter some basic notions are introduced. The second chapter is a diverse collection of real-life situations where a first order differential equation pops up in the mathematical modelling. Each problem was chosen with a specific purpose in mind. In general it is shown that there is much more to these mathematical models than merely solving differential equations; on the contrary, in some cases it is not even necessary to solve the differential equation.

In the third chapter some elementary numerical methods are discussed, with the accent on ways to develop more accurate methods rather than presenting fairly sophisticated numerical methods. In a real-life situation the applied mathematician will probably use an efficient standard numerical program which is part of the software of his computer.

Laplace transforms are introduced in the fourth chapter as a very necessary tool for solving systems of linear differential equations. In the next chapter we look at mathematical models in which systems of differential equations appear. Again each problem is chosen to highlight different aspects of mathematical models and/or the theory of differential equations.

In the sixth chapter the standard applications of second order differential equations in mechanical vibrations and electrical networks are discussed, but it also includes a mathematical model on the ignition of a car which teaches us to think before we spend a lot of time solving differential equations. In the final chapter two situations are modelled, namely, the pendulum and competing species, to illustrate qualitative solutions.

Through the years many people have contibuted to the creation of this book. The discussions with my colleagues were very helpful, in particular the ideas of Gerhard Geldenhuys and Philip Fourie. The suggestion to write this book came from Alan Jeffrey of the University of Newcastle on Tyne. His interest and comments are deeply appreciated. The encouragement of Navin Sullivan in London carried the project through. Finally, without the help of Hester Uys and Jan van Vuuren on the computer, and Christelle Goldie with the sketches, there would not be a book at all.

Although this book may look like a jumble of disconnected applications, be warned - there is method in the madness, and you may learn more than you thought you would. So start right away and enjoy the book.

TP Dreyer

1

Introduction

1.1 Mathematical Modelling

Although mathematics had already been applied to real-life problems by the Egyptians and other ancient civilizations, the term "mathematical model" is a fairly recent addition to the mathematical vocabulary. The term signifies an attempt to describe the interplay between the physical world on the one hand and abstract mathematics on the other hand. It is customary to refer to a collection of equations, inequalities, and assumptions as the "model", but the term "mathematical modelling" means more than that: it is an orderly structured manner in which a real-life problem might be tackled.

The process of mathematical modelling can be broken up into seven different stages, as is shown in the flow chart in Figure 1.1.1.

(1) **Identification:** The questions to be answered must be clarified. The underlying mechanism at work in the physical situation must be identified as accurately as possible. Formulate the problem in words, and document the relevant data.

(2) **Assumptions:** The problem must be analysed to decide which factors are important and which factors are to be ignored so that realistic assumptions can be made.

(3) **Construction:** This is the translation of the problem into mathematical language which normally results in a collection of equations and/or inequalities after the variables had been identified. The "word" problem is transformed into an abstract mathematical problem.

(4) **Analysis:** The mathematical problem is solved so that the unknown variables are expressed in terms of known quantities, and/or it is analysed to obtain information about parameters.

(5) **Interpretation:** The solution to the abstract mathematical problem must be compared to the original "word" problem to see if it makes sense in the real-world situation. If not, go back to formulate more realistic assumptions, and construct a new model.

(6) **Validation:** Check whether the solution agrees with the data of the real-world problem. If the correlation is unsatisfactory, return to the "word" problem for a re-appraisal of the data and the assumptions. Modify or add assumptions and construct a new model.

(7) **Implementation:** If the solution agrees with the data, then the model can be used to predict what will happen in the future, or conclusions can be drawn to help in future planning, etc. In the case of predictions care should be taken to determine the time interval in which the predictions are valid.

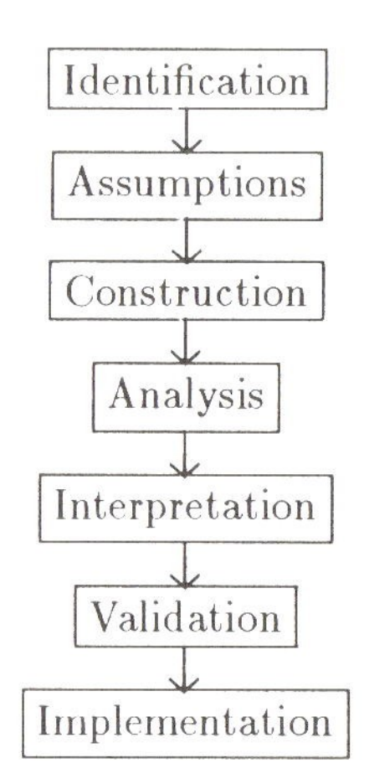

Figure 1.1.1: **Flow chart of the modelling process**

In a specific problem we may not use all seven stages, or some stages may be trivial. However, even though we will not state each stage explicitly in every model in this book, it is always the way in which we think about each problem.

Obviously a single book cannot cover all the different types of mathematical models. To narrow the field one could differentiate models either by the type of physical problem or by the type of mathematics needed to solve the problem. For example, in the first case only problems related to epidemics could be discussed, or in the second case the models can be restricted to those using systems of linear equations. In this book the second option was chosen with ordinary differential equations as the unifying theme of the different mathematical models.

- Do: Exercise 1 in §1.5.

1.2 Boundary Value Problems

Ordinary differential equations originate naturally in most of the branches of science. In this book we shall look at examples in diverse fields like physics, engineering, biology, economics, and medicine. A thorough knowledge of differential equations is a powerful tool in the hands of an applied mathematician with which to tackle these problems. We shall try to equip you with this tool in the pages that follow. But first a few general definitions and background material.

Definition 1.1

An ordinary differential equation *is an equation in which an unknown function $y(x)$ and the derivatives of $y(x)$ with respect to x appear. If the n-th derivative of y is the highest derivative in the equation, we say that the differential equation is of the* n-th order. *The* domain *of the differential equation is the set of values of x on which $y(x)$ and its derivatives are defined.*

Usually the domain is an interval I on the real line where I could be a finite interval (a, b), the positive real numbers $(0, \infty)$, the non-negative real numbers $[0, \infty)$, or even the whole real line $(-\infty, \infty)$. We shall use this notation: a square bracket denotes that the endpoint is included and a round bracket that the endpoint is not included in the interval.

Definition 1.2

If the differential equation is linear in the dependent variable y and all its derivatives, we refer to it as a linear differential equation. *If not, we shall call it a* nonlinear differential equation. *The general form of a linear differential equation of the n-th order is*

$$a_n(x)\frac{d^n y}{dx^n} + \ldots + a_1(x)\frac{dy}{dx} + a_0(x)y = b(x) \quad (1.2.1)$$

The main purpose of this book is to teach you how to construct *mathematical models* of selected real-life problems. These problems were chosen to include an ordinary differential equation in the model. To obtain a meaningful answer to the given problem, the differential equation must be solved. The solution must then be interpreted in the light of the assumptions that were made in the construction of the model, as we have seen in §1.1.

Apart from the differential equation, the mathematical model will typically also include prescribed values of y and/or the derivatives of y at isolated points in the interval I, usually one or both of the endpoints.

Definition 1.3

The prescribed values of y and/or the derivatives of y at the endpoints of the interval are called boundary values, *and the problem of finding $y(x)$ where y appears in a differential equation as well as in prescribed boundary values is called a* boundary value problem. *If the independent variable x represents time and all the boundary values are specified at the left endpoint of the interval, the boundary value problem is called an* initial value problem.

When a boundary value problem is encountered, there are four questions which must be answered:

(1) What is meant by a "solution"?

(2) Does a solution exist?

(3) Is there more than one solution?

(4) Is the solution a continuous function of the boundary values?

Let us briefly discuss each of these important questions.

There are different ways in which "solution" can be understood. For example, do we require that the function $y(x)$ should satisfy the differential equation everywhere in I, or could there be a few exceptional points where the highest derivative does not exist. Let $i(I)$ denote the interior of I which means the largest open interval in I. If I is itself an open interval, then of course $i(I) = I$; otherwise one or both the endpoints will be excluded.

In this book, unless specifically stated otherwise, we shall use the word "solution" in the following sense:

Definition 1.4

A solution *of a given boundary value problem on the interval I is a function which is continuous on I, satisfies the differential equation at every point $x \in i(I)$, and agrees with the prescribed boundary values.*

Note that the definition implies that the derivatives which appear in the equation must exist everywhere in i(I). A function which is differentiable on an open interval is also continuous there (see [16] p. 166), but the converse is not true – for example $f(x) = |x|$ is continuous on $[-1; 1]$, but the derivative does not exist at $x = 0$, even though the left hand derivative and the right hand derivative do exist. (Remember that a function is continuous at a point $x = a$ if the left and right hand limits exist and are both equal to $f(a)$; and similarly for a derivative, both left and right hand limits must exist and be equal to each other.)

However, the necessity of a more general definition of the term "solution" will be shown in §2.8. There we shall need a slightly weaker condition than continuity, namely piecewise continuity. We shall also use piecewise continuous functions in §4.2.

Definition 1.5

A function f is said to be piecewise continuous *on a finite interval $[a, b]$ if the interval can be subdivided into a finite number of intervals with f continuous on each of these intervals and if the jump in the value of f at each of the endpoints of these intervals is finite.*

Note that the jump in the value of f is the difference between the left and right hand limits, which must both exist if the jump is to be finite. We shall continue the discussion of piecewise continuity in §2.8 and §4.2.

- Do: Exercise 2 in §1.5.

The *existence* of a solution is usually settled by finding the solution explicitly with the aid of mathematical techniques. However, there are many differential equations whose solutions cannot be expressed in terms of elementary functions (For example, the pendulum equation $\theta'' + \omega^2 \sin\theta = 0$ – see §7.2.) Then numerical techniques must be utilized to calculate the value of the solution approximately at selected points in the interval I. These calculations only make sense if one knows beforehand that a solution does indeed exist. In these cases the existence of a solution to the boundary value problem must be proven indirectly, without knowing what the solution really looks like. The construction of proofs for the so-called existence theorems is an important research area in mathematics.

Once the question of the existence of a solution is settled, the next logical question is whether the known solution is the only possible solution. If this is the case, then we say the solution is *unique*. Clearly the uniqueness of the solution is very important for the interpretation of the results of the model – in fact, if the solution is not unique, then all the possible solutions must be found before any meaningful conclusions can be drawn. Sometimes the uniqueness follows immediately by the manner in which the solution was found; in other cases it is more complicated and one usually relies on special theorems which were proved beforehand. Both these approaches will be encountered in this book.

Finally, it is also very important to know how sensitive the solution is to changes in the boundary values. The main reason for this is that they are subject to errors. The crucial question is whether a small error in the boundary values will cause a small error in the solution. If this is the case the solution must be a continuous function of the boundary values.

- Do: Exercises 3, 4, 5 in §1.5.

1.3 Direction Fields

Consider the initial value problem

$$\frac{dy}{dx} = F(x,y), \qquad y(0) = \alpha \tag{1.3.1}$$

where α is a prescribed initial value and F is a given function. If a solution $y = f(x)$ exists, then the graph of this solution is a curve in the (x, y)–plane passing through the point $(0, \alpha)$ on the y– axis. For different values of α, different solution curves are obtained so that one can think of the (x, y)–plane as being filled with solution curves in the sense that through every point in the plane passes a solution curve corresponding to some value of α (provided, of course, that F is defined and smooth enough). The slope of the tangent to each solution curve at any point $(a, f(a))$ on the curve is $F(a, f(a))$ by (1.3.1). Consider, for example, the boundary value problem

$$\frac{dy}{dx} = \frac{2x\left(x^2+1\right)}{y} \tag{1.3.2}$$

$$y(0) = 1 \tag{1.3.3}$$

We shall see in Chapter 2 that the solution is $y = x^2 + 1$ (see Exercise 32 in §2.12).

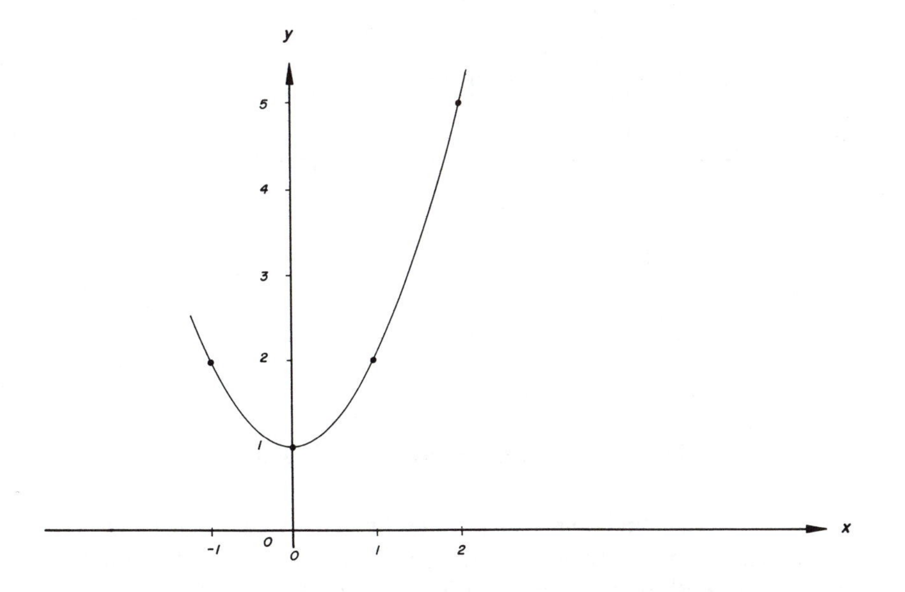

Figure 1.3.1: **Solution of (1.3.2) and (1.3.3)**

By differentiation it follows that the slope of this solution curve is $2x$. On the other hand, by (1.3.2) we have

$$\frac{dy}{dx} = \frac{2x\left(x^2+1\right)}{x^2+1} = 2x.$$

The function $F(x, y)$ in (1.3.1) prescribes at each point (x, y) in the plane where F is defined, a slope (or direction) for the solution curve which passes through (x, y). We say briefly that the differential equation prescribes a *direction field*. If we choose a suitable rectangular mesh in the plane, a useful picture of the direction field can be obtained. At each point (a, b) of the mesh in the $(x, y)-$plane, a small line segment (also called a lineal element) is drawn from (a, b) with slope $F(a, b)$. Since each of these line segments is a tangent to a solution curve, the picture of all these line segments gives an indication of the shape of the solution curves. Obviously, the finer the rectangular mesh, the better this indication will be – provided that the mesh is not finer than the thickness of the line segments!

In this way an idea of the behaviour of the solution curves can be obtained, even though the solution of (1.3.1) may not be known.

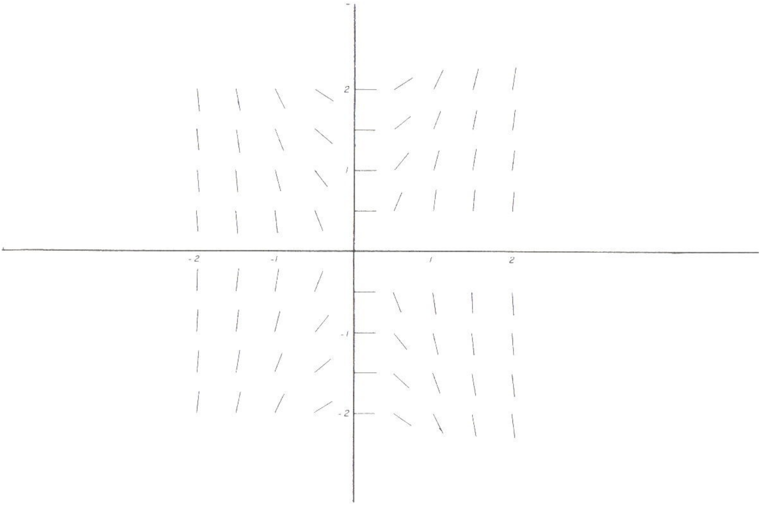

Figure 1.3.2: **Direction field of (1.3.2)**

Let us draw the direction field for the differential equation in (1.3.2) on the subset $R = \{(x, y) : \ -2 \le x \le 2, \ -2 \le y \le 2\}$ with the rectangular mesh points at ± 2, ± 1.5, ± 1.0, ± 0.5, and 0 for both x and y. Then we need the slope at 81 mesh points. At the origin the slope is indeterminate, on the remainder of the x−axis it is not defined, and on the remainder of the y−axis it is zero. Note, however, that we need only calculate the slope at the 16 mesh points in the first quadrant because of symmetry in the expression for the slope in (1.3.2). For example, by (1.3.2) the slope at $(1, 2)$ is 2; then it follows immediately that the slope at $(-1, 2)$ must be -2, at $(1, -2)$ must be -2, and at $(-1, -2)$ must be 2. At each mesh point draw a line segment of length 0.25 with the corresponding slope to the right of the mesh point, as is shown in Figure 1.3.2. Note how the solution curve in Figure 1.3.1 fits into this pattern of directions.

The picture in Figure 1.3.2 is rather crude, even though it took some time and effort. Let us now utilize the computer to do the hard work. For this we need an *algorithm*.

Definition 1.6

An algorithm *is a set of instructions which must be implemented consecutively to obtain the desired result.*

In a way it is just the mathematical equivalent of a recipe for baking a cake. We shall devise an algorithm for the general case (1.3.1) with an arbitrary mesh size h, which we can keep as a handy tool in the future (unless the reader already has a good graphics package available). To identify points where the slope is undefined, let us assume that the function F can be written as a quotient

$$F(x, y) = \frac{N(x, y)}{D(x, y)}$$

where N and D are defined (and finite) for all the points in the subset of the plane under consideration.

Algorithm: DIRFIELD

(1) Specify the ranges $[a, b]$ on the x−axis and $[c, d]$ on the y−axis (a, b, c, and d are integers).

(2) Draw and label the axes with integer subdivisions.

(3) Specify the mesh size h.

(4) Define the functions $N(x, y)$ and $D(x, y)$.

(5) Let $x = a$.

(6) Let $y = c$.

(7) Calculate $D(x, y)$. If the result is zero, go to step 10; otherwise continue to the next step.

(8) Calculate $F(x, y) = \frac{N(x,y)}{D(x,y)}$.

(9) Draw the line segment from (x, y) with slope $F(x, y)$ and length $\frac{h}{2}$.

(10) $y = y + h$.

(11) If $y < d$, go to step 7; otherwise continue to the next step.

(12) $x = x + h$.

(13) If $x < b$, go to step 6; otherwise continue to the next step.

(14) End.

This algorithm must now be translated into a *program* for the computer, according to the computer language of the reader's choice. For example, in TURBO.PASCAL 4.0 a program would look like this:

```
program DIRFIELD;
uses
        Graph;
var
        GraphDriver: integer;
        GraphMode: integer;
        A,B,C,D,I,M,N,P,Q,S,T,V,W: integer;
        x,y,h,U,L: real;
function NUM(x,y: real): real;
        begin
        NUM:= 2*x*(x*x + 1);
        end;
function DEN(x,y: real): real;
        begin
        DEN:= y
        end;
begin
        Writeln('DIRFIELD');
        Writeln('Enter the intervals on the axes:');
        Writeln('(Numbers must be integers!)');
        Writeln('Left hand (non-positive) endpoint on x-axis=: A =');
        Read(A);
        Writeln('Right hand (non-negative) endpoint on x-axis=: B =');
        Read(B);
        If A*A + B*B = 0 then Writeln(' x-interval zero!!');
        Writeln('Lower (non-positive) endpoint on the y-axis: C =');
        Read (C);
        Writeln('Upper (non-negative) endpoint on the y-axis: D =');
        Read(D);
```

```
        If C*C + D*D = 0 then Writeln('y-interval zero!!');
        Writeln('Enter the mesh size:');
        Writeln('h =');
        Read(h);
        GraphDriver:= Detect;
        InitGraph(GraphDriver,GraphMode,'');
        M:= ROUND(10 - 499*A/(B - A));
        MoveTo(M,40);
        LineTo(M,189); (*Draw the y-axis*)
        N:= ROUND(189 + 149*C/(D - C));
        MoveTo(10,N);
        LineTo(509,N); (*Draw the x-axis*)
        For I:=0 to B - A do (*Marking of integer points on x axis*)
            begin
            P:= ROUND(10 + 499*I/(B - A));
            MoveTo(P,N);
            LineTo(P,N + 3);
            end;
        For I:=0 to D - C do (*Marking of integer points on y axis*)
            begin
            Q:= ROUND(189 - 149*I/(D - C));
            MoveTo(M -5,Q);
            LineTo(M,Q);
            end;
        x:=A;
        y:=C;
        While x<=B do
            begin
            While y<=D do
                begin
                U:= NUM(x,y);
                L:= DEN(x,y);
                If ABS(L)>0.0001 then
                    begin
                    S:= ROUND (10 + 499*(x - A)/(B - A));
                    T:= ROUND (189 - 149*(y - C)/(D - C));
                    V:= ROUND (10 + (x + h*COS(ARC-
                       TAN(U/L))/2 - A)*499/(B - A));
                    W:= ROUND (189 - (y + h*SIN(ARC-
                       TAN(U/L))/2 - C)*149/(D - C));
                    MoveTo(S,T);
                    LineTo(V,W);
                    end;
                y:= y + h;
                end
            y:= C;
            x:= x + h;
            end;
        Halt;
        CloseGraph;
    end.(Dirfield)
```

The labels of the axes were omitted (on purpose to shorten the program), and the notation F in step 8 of the algorithm was not used in the program. Note that if the denominator D is zero at a mesh point, then the line segment is not drawn there. Hence the omission of a line segment in the mesh implies that the slope is undefined at that particular mesh point.

- Do: Exercise 6 in §1.5.

Note: In the Exercises 33 and 34 in §2.11 you will be required to draw graphs of functions of the type $y = f(x)$. There are many good programs available to do this, and if you are familiar with such a program on your computer, skip the rest of this section. If you do not have the type of program at hand, we can quickly devise a simple program to bail you out:

Algorithm: OURGRAPH

(1) Specify the ranges $[a, b]$ on the x–axis and $[c, d]$ on the y–axis. (a, b, c, and d are integers.)

(2) Draw and label the axes with integer subdivisions.

(3) Specify the mesh size h.

(4) Define the function $f(x)$.

(5) Let $x = a$.

(6) Draw the line segment from $(x, f(x))$ to $(x + h, f(x + h))$.

(7) $x = x + h$.

(8) If $x \leq b$, go to step 6; otherwise continue to the next step.

(9) End.

Translated into TURBO-PASCAL 4.0 a program to draw the graph of $y = \sqrt{2\cos x + 1}$ would look like this:

```
program OURGRAPH;
uses
        Graph;
var
        GraphDriver: integer;
        GraphMode: integer;
        A,B,C,D,I,M,N,P,Q,S,T,V: integer;
        x,y,h: real;
function FUNC(x: real): real;
        begin
        FUNC:= SQRT(2*(cos(x) + 1));
        end;
begin
        Writeln('OURGRAPH');
        Writeln('Enter the intervals on the axes:');
        Writeln('(Numbers must be integers!)');
        Writeln('Left hand (non-positive) endpoint on x-axis: A =');
        Read(A);
```

```
        Writeln('Right hand (non-negative) endpoint on x-axis: B =');
        Read(B);
        If A*A + B*B = 0 then Writeln(' x-interval zero!!');
        Writeln('Lower (non-positive) endpoint on the y-axis: C =');
        Read (C);
        Writeln('Upper (non-negative) endpoint on the y-axis: D =');
        Read(D);
        If C*C + D*D = 0 then Writeln(' y-interval zero!!');
        Writeln('Enter the mesh size:');
        Writeln('h =');
        Read(h);
        GraphDriver:= Detect;
        InitGraph(GraphDriver,GraphMode,");
        M:= ROUND(10 - 499*A/(B - A));
        MoveTo(M,40);
        LineTo(M,189,); (*Draw the y-axis*)
        N:= ROUND(189 + 149*C/(D - C));
        MoveTo(10,N);
        LineTo(509,N); (*Draw the x-axis*)
        For I:=0 to B - A do (*Marking of integer points on x-axis*)
              begin
              P:= ROUND(10 + 499*I/(B - A));
              MoveTo(P,N);
              LineTo(P,N + 3);
              end;
        For I:=0 to D - C do (*Marking of integer points on y-axis*)
              begin
              Q:= ROUND(189 - 149*I/(D - C));
              MoveTo(M -5,Q);
              LineTo(M,Q);
              end;
        x:= A;
        y:= FUNC(A);
        V:= ROUND(189 -149*(y - C)/(D -C));
        MoveTo (10,V);
        While x<=(B -h) do
              begin
              S:= ROUND (10 + 499*(x + h - A)/(B - A));
              Y:= FUNC(x + h);
              T:= ROUND (189 - 149*(y - C)/(D - C));
              LineTo(S,T);
              x:= x + h;
              end;
        Halt;
        CloseGraph;
end.(Ourgraph)
```

1.4 Finding Zeros

In many of the applications in the rest of the book the following problem crops up: find the zeros of the function $f(x)$ on the interval $[a,b]$. In other words, find the numbers $k_1,\ k_2,\ldots,\ k_n$ such that $f(k_i) = 0$ and $a \leq k_i \leq b$ for $i = 1,2,\ldots,n$. If the problem cannot be solved by an analytical method, we must fall back on a numerical procedure. In this section we briefly discuss two such procedures.

First of all we must determine a subinterval of $[a,b]$ in which a specific zero $x = k$ lies. This is usually done by drawing a qualitative graph (in other words, a graph drawn by hand which shows the general behaviour of the function rather than numerical values at specific points). Then this subinterval is used to obtain initial approximations from which better approximations of the zero are derived.

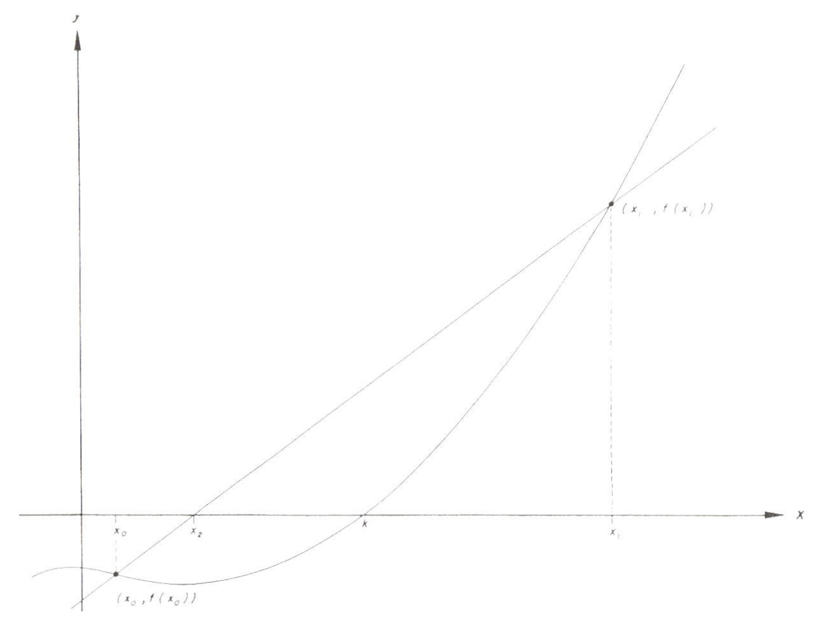

Figure 1.4.1: **Secant method**

Secant Method

We shall denote the successive approximations of the zero $x = k$ by $x_0,\ x_1,\ x_2,\ \ldots$ Suppose x_0 and x_1 are obtained from the subinterval in which the zero lies. Draw the secant through $(x_0,\ f(x_0))$ and $(x_1,\ f(x_1))$. Let $x = x_2$ be the point of intersection between the secant and the x–axis. Then x_2 is the next approximation. To calculate x_2, we need the equation for the secant

$$\frac{y - f(x_1)}{f(x_0) - f(x_1)} = \frac{x - x_1}{x_0 - x_1} \tag{1.4.1}$$

and then substitute $y = 0$ in (1.4.1)

$$x_2 = x_1 - f(x_1)\frac{x_1 - x_0}{f(x_1) - f(x_0)}$$

We then repeat the process with x_1 and x_2 (instead of x_0 and x_1) to obtain x_3, etc. In general with x_{n-1} and x_n, we obtain

$$x_{n+1} = x_n - f(x_n)\frac{x_n - x_{n-1}}{f(x_n) - f(x_{n-1})}, \qquad n = 2, 3, \ldots \qquad (1.4.2)$$

This is known as the secant method. This method can be used to calculate k to a prescribed accuracy, called the tolerance.

Algorithm: SECANT

(1) Define the function $f(x)$.

(2) Specify the initial approximation x_0 and x_1, the tolerance E, and the maximum number of iterations N.

(3) Calculate $f_0 = f(x_0)$ and $f_1 = f(x_1)$.

(4) Set $I = 2$.

(5) Calculate $X = x_1 - f_1\frac{(x_1 - x_0)}{(f_1 - f_0)}$.

(6) Print X.

(7) If $|X - x_1| < E$ then stop; otherwise continue to the next step.

(8) If $I = N + 1$ then print failure message and stop; otherwise continue to the next step.

(9) Set $x_0 = x_1$, $f_0 = f_1$, $x_1 = X$, $f_1 = f(X)$, and $I = I + 1$.

(10) Go to step 5.

(11) END.

Note that (1.4.2) must not be simplified to the form

$$x_{n+1} = \frac{x_{n-1}f(x_n) - x_n f(x_{n-1})}{f(x_n) - f(x_{n-1})} \qquad (1.4.3)$$

because if x_n is close to x_{n-1} cancellation may occur. Furthermore, (1.4.2) shows clearly in the second term on the right hand side the amount by which x_n will be corrected to obtain x_{n+1}.

Consider the problem to determine the zero of

$$e^{-x} - x = 0. \qquad (1.4.4)$$

In TURBO-PASCAL 4.0 a program for the secant method looks like this:

```
program SECANT;
var
       I,N: integer;
       X0,X1,E,F0,F1,X:real;
function F(x: real):real;
       begin
       F:= EXP(-x) - x
       end;

begin
       Writeln('SECANT METHOD');
       Writeln('The tolerance is');
       Writeln('E =');
       Read(E);
       Writeln('The maximum number of iterations allowed is');
       Writeln('N =');
       Read(N);
       Writeln('The first two approximations are');
       Writeln('X0 =');
       Read(X0);
       Writeln('X1 =');
       read(X1);
       Writeln('The successive approximations are');
       F0:= F(X0);
       F1:= F(X1);
       I:= 2;
       While I<=N+1 do
              begin
              X:= X1 - F1*(X1 - X0)/(F1 - F0);
              Writeln('x = ',X);
              If ABS(X - X1)<E then halt;
              X0:= X1;
              F0:= F1;
              X1:= X;
              F1:=F(X);
              I:= I + 1;
              end;
       Writeln;
       Writeln('Required accuracy not obtained!!!!');
       Writeln('Either try other initial approximations X0 and X1');
       Writeln('or increase N and/or E');
end.(Secant)
```

Note that the function f must be defined in the program.

The secant method will always converge if f' (the derivative of f) is nonzero and f'' (the second derivative of f) is continuous on an interval containing the zero $x = k$. It can also be shown (see [17] p. 220) that

$$|k - x_{n+1}| \approx C|k - x_n|^p \tag{1.4.5}$$

where $p = \frac{1+\sqrt{5}}{2}$ and

$$C^p = \frac{|f''(k)|}{2|f'(k)|}$$

- Do: Exercise 7 in §1.5.

Newton-Raphson Method

Let us denote the successive approximations of the zero $x = k$ again by x_0, x_1, ... Let x_0 be an initial approximation. At the point $(x_0, f(x_0))$ draw a tangent to intersect the x−axis at $x = x_1$, which is then the next approximation of the zero.

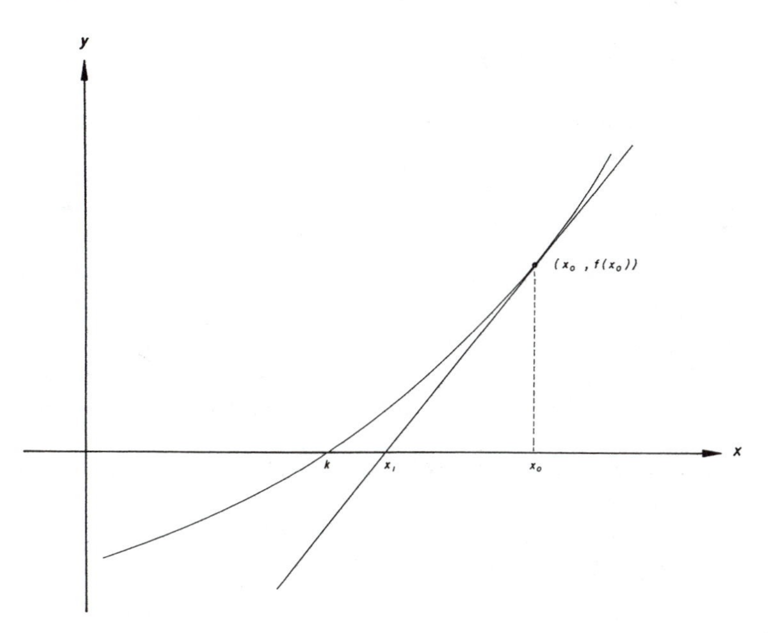

Figure 1.4.2: **Newton-Raphson method**

To calculate x_1, we need the equation of the tangent:

$$y - f(x_0) = f'(x_0)(x - x_0) \tag{1.4.6}$$

where f' denotes the derivative of f. Substitute $y = 0$ to obtain x_1:

$$x_1 = x_0 - \frac{f(x_0)}{f'(x_0)}$$

Repeat now the process with x_1 instead of x_0 to obtain x_2. In general

$$x_{n+1} = x_n - \frac{f(x_n)}{f'(x_n)}, \qquad x = 0, 1, 2, \ldots \tag{1.4.7}$$

This is known as the Newton-Raphson method. Note that if $|f'(x_n)|$ is small relative to $|f(x_n)|$, the method will not be efficient.

Algorithm: NEWTON

(1) Define the function $f(x)$.

(2) Specify the initial approximation x_0, the tolerance E, and the maximum number of iterations N.

(3) Calculate $f_0 = f(x_0)$ and $f_0' = f'(x_0)$.

(4) Check the value of $|f_0'|$; if it is small, print a warning and if it is zero, stop.

(5) Set $I = 1$.

(6) Calculate $X = x_0 - \frac{f_0}{f_0'}$.

(7) Print X.

(8) If $|X - x_0| < E$ then stop; otherwise continue to the next step.

(9) If $I = N$ then print failure message and stop; otherwise continue to the next step.

(10) Set $x_0 = X$, $f_0 = f(X)$, $f_0' = f'(X)$, and $I = I + 1$.

(11) Go to step 6.

(12) END.

In TURBO-PASCAL 4.0 a program for the Newton-Raphson method to solve (1.4.5) looks like this:

```
program NEWTON;
var
        I,N: integer;
        X0,E,F0,DF0,X:real;
function F(x: real):real;
        begin
        F:= EXP(-x) - x
        end;
function DF(x: real): real;
        begin
        DF:= -EXP(-x) - 1
        end
```

```
begin
      Writeln('NEWTON-RAPHSON METHOD');
      Writeln('The tolerance is');
      Writeln('E =');
      Read(E);
      Writeln('The maximum number of iterations allowed is');
      Writeln('N =');
      Read(N);
      Writeln('The first approximation is');
      Writeln('X0 =');
      Read(X0);
      Writeln('The successive approximations are');
      F0:= F(X0);
      DF0:= DF(X0);
      I:= 1;
      WhileI<=N do
            begin
            If ABS(DF0)<0.5 then writeln('Derivative is',
               DF0,'<0.5');
            If DF0 = 0 then halt;
            X:= X0 - F0/DF0;
            Writeln('x = ',X);
            If ABS(X - X0)<E then halt;
            X0:= X;
            F0:= F(X);
            DF0:= DF(X);
            I:= I + 1;
            end;
      Writeln;
      Writeln('Required accuracy not obtained!!!!');
      Writeln('Either try other initial approximation X0');
      Writeln('or increase N and/or E');
end.(Newton)
```

Note that f and f' must be defined in the program. The Newton-Raphson method converges faster than the secant method, but if f' is complicated, the secant method may be preferable. It can be shown (see [17] p. 224) that the Newton-Raphson method converges if $f'(k) \neq 0$ and f'' (the second derivative of f) is continuous on an interval containing the zero $x = k$. Furthermore

$$|k - x_{n+1}| \approx C|k - x_n|^2,$$

where

$$C = \frac{|f''(k)|}{\alpha|f'(k)|}$$

- Do: Exercise 8 in §1.5.

1.5 Exercises

(1) You are conducting an experiment on the population dynamics of the common housefly. A number of flies are kept in a very large bottle with a fixed amount of daily food. The number of flies alive in the bottle is counted every morning at 9:00. After a few days the experimental results look like this:

Time (in days)	Flies
0	200
1	270
2	360
3	470
4	590
5	700

You would like to predict the population at some future date.

(a) Denote the population by X and the time by t and assume that $X(t)$ is a continuous function for $t \geq 0$. (Is this a reasonable assumption?) Assume also as a wild guess that the population increases daily by the same constant number k, where k is some fixed integer. (If you are worried about this assumption, read the comments on assumptions in [45] on p. 48!!) Construct the mathematical model and show that X is a linear function of t. Draw a graph of the function X and also plot the data in the table above. (Hint: If you have a derivative in your model, you are on the wrong track - this is a simple problem!)

(b) Rewrite the problem in blocks as in Figure 1.1.1 with the relevant information in each block to show how the process of mathematical modelling was carried out. How do you interpret your solution for large values of t?

(c) You should find that the validation is not satisfactory. Try to determine the best possible fit of the data by shifting a ruler so that the resulting straight line is as close as possible to the data (that is, the sum of the differences between the line and each data point should be as small as possible - what about positive and negative differences?) Is your best possible fit satisfactory?

(d) Try a new assumption for the growth of the population - something like: the population increases daily by the same constant fraction a of the population the previous day. Solve this model, interpret the solution, and validate on a graph. Are you now happy to proceed to implementation? (Hint: You should not be!)

(e) Stop wasting any more time on this problem, since you clearly need some new tools to handle it. Now you are really motivated to read this book!

(2) Show that

(a) the function

$$f(x) = \begin{cases} 3e^{2x}, & -1 \le x < 0 \\ 3e^{x}, & 0 \le x \le 1 \end{cases}$$

is a continuous function on the interval $[-1; 1]$;

(b) the derivative of f is a piecewise continuous function on $[-1; 1]$;

(c) the derivative of f does not exist at $x = 0$;

(d) the function f satisfies the differential equation

$$\frac{dy}{dx} - p(x)y = 0, \qquad p(x) = \begin{cases} 2, & -1 \le x < 0 \\ 1, & 0 < x \le 1 \end{cases}$$

everywhere on the open interval $(-1; 1)$ except at $x = 0$.

(3) For each of the following, state (i) the order of the differential equation, (ii) the largest domain (interval) where it is defined, and (iii) whether it is linear or nonlinear:

(a) $\frac{d^3y}{dx^3} + 6x^2\frac{dy}{dx} - 20e^{2x}\frac{d^4y}{dx^4} = \sin x$

(b) $\left(\frac{dy}{dx}\right)^2 - xy^2 = 0$

(4) Check whether the following functions are solutions of the given boundary value problem:

(a) $y = \sin 2x$; $\frac{d^2y}{dx^2} + 4y = 0$, $y(0) = 0$, $y(\pi) = 1$

(b) $y = |x|$; $\left(\frac{dy}{dx}\right)^2 = 1$, $y(-1) = 1$, $y(1) = 1$

(c) $y = \frac{1}{2-e^{-t}}$; $\frac{dy}{dt} = y - 2y^2$, $y(0) = 0$

(d) $y = \frac{1}{2}$; $\frac{dy}{dx} = y - 2y^2$, $y(0) = \frac{1}{2}$

(5) Show that the boundary value problem

$$\frac{dy}{dx} = 2\sqrt{y}, \qquad y(0) = 0$$

has two solutions, namely $y = x^2$ $(0 \le x < \infty)$ and $y = 0$ $(0 \le x < \infty)$.

(6) Use the program DIRFIELD (or any other similar program) to draw

(a) the direction field of (1.3.2) as in Figure 1.3.2 with $h = 0.5$

(b) the direction field of (1.3.2) with $h = 0.1$

(c) the direction field of

$$\frac{dy}{dx} = -\frac{\sin x}{y}$$

for $0 \leq x \leq 6$ and $0 \leq y \leq 4$ with $h = 0.25$

(d) the direction field of

$$\frac{dy}{dx} = -\frac{\sin x}{y} - 1$$

for $0 \leq x \leq 6$ and $0 \leq y \leq 4$ with $h = 0.25$

(We shall continue this exercise in Exercise 34 in §2.12.)

(7) Use the secant method to find the zeros of the following functions $f(x)$ on the interval $[0, 1]$ correct to 3 decimal places.

(a) $f(x) = x^3 - 7x + 4$

(b) $f(x) = e^{-x} - 2\sin x$

(c) $f(x) = xe^x \cos x - 1$

(8) Do Exercise 7 with the Newton-Raphson method instead of the secant method.

2

First Order Differential Equations

2.1 Introduction

In this chapter we shall look at mathematical models in which a first order differential equation plays a role. The equation shall always be in normal form

$$\frac{dy}{dx} = F(x, y) \tag{2.1.1}$$

If the function F is linear in y,

$$F(x, y) = p(x)\ y + q(x) \tag{2.1.2}$$

where p and q are functions of x only, we say that the differential equation (2.1.1) is linear; otherwise it is nonlinear.

The property of linearity plays an important role in the analysis of a differential equation. For example, if (2.1.1) is linear and $y = f(x)$ satisfies the equation,

$$\frac{dy}{dx} + p(x)\ y = 0 \tag{2.1.3}$$

then $y = cf(x)$ where c is an arbitrary constant, also satisfies (2.1.3) because

$$\frac{d}{dx}(cf(x)) = c\frac{df}{dx} = c[-p(x)f(x)] = -p(x)[cf(x)]$$

For the linear differential equation

$$\frac{dy}{dx} + p(x)\ y = q(x) \tag{2.1.4}$$

we call $cf(x)$ the complementary function, and any function $g(x)$ which satisfies (2.1.4) a particular integral.

Note that

$$y = cf(x) + g(x) \tag{2.1.5}$$

satisfies (2.1.4) for an arbitrary constant c.

It was shown in §1.3 that different boundary values correspond to different solution curves in the (x, y)-plane. Given the boundary value

$$y(0) = \alpha \tag{2.1.6}$$

we can now match the constant in (2.1.5) to this boundary value by substitution:

$$\alpha = cf(0) + g(0).$$

Hence the value of c can be found, provided that $f(0) \neq 0$. Should $f(0) = 0$, there would exist a solution for only one boundary value, namely $\alpha = g(0)$. However, we shall show below that $f(0) \neq 0$ so that a solution exists for any prescribed boundary value α. In general, a solution of (2.1.1) and (2.1.6) may not exist. For example

$$\frac{dy}{dx} = -y\sqrt{y^2 - 4}, \qquad y(0) = 1 \tag{2.1.7}$$

has no solution, because the slope is only defined when $|y| \geq 2$, and the boundary value requires that $y = 1$. It was also shown in Exercise 5 of §1.5 that there could be more than one solution. We shall discuss theorems which guarantee the existence and uniqueness of the solution of (2.1.1) and (2.1.6) in §3.1 and §2.3.

There are several methods to find the solution of (2.1.1) explicitly, depending on the form of the function $F(x, y)$. We shall concentrate on four main types, and refer the reader to books like [9], [10], [50], and [62] for more specialized information on other types.

Fundamental Theorem of Calculus

The simplest type of differential equation is that when $F(x, y)$ is a function of x only:

$$\frac{dy}{dx} = F(x), \qquad y(a) = c. \tag{2.1.8}$$

The solutions of such differential equations are described by the *fundamental theorem of calculus*.

Theorem 2.1

If the function $F(x)$ is continuous on the closed interval $[a,b]$, then there is one and only one solution $y = f(x)$ of the boundary value problem (2.1.8). This solution is

$$f(x) = c + \int_a^x F(t)\, dt$$

□

We shall not prove this well-known theorem (see, for example, [16] p. 188).

Separable Equations

Definition 2.1

Any equation which can be written in the form

$$\frac{dy}{dx} = \frac{f(x)}{g(y)} \tag{2.1.9}$$

is called a separable equation. □

If we multiply both sides of (2.1.9) by $g(y)$ and integrate, we note that the integral on the left hand side can be written as

$$\int g(y)\frac{dy}{dx}\, dx = \int g(y)\, dy$$

by the standard theorem on substitutions in an integral. (See, for example, [16] p. 265). Hence, the variables can be separated by integration

$$\int g(y)\, dy = \int f(x)\, dx + c \tag{2.1.10}$$

where c is an arbitrary constant. Provided that the two integrals can be found, the solution of equation (2.1.9) can be found at least in an implicit form.

- Do: Exercises 32, 33, 34 in §2.12.

Linear Equations

As was discussed above, (2.1.1) reduces in this case to

$$\frac{dy}{dx} + p(x)\ y = q(x) \tag{2.1.11}$$

The main obstacle here is that of the two terms on the left hand side. If one could somehow write the left hand side as the derivative of a single function, we need merely integrate both sides by Theorem 2.1 above. The one obvious way in which the derivative of a function can produce two terms is when the product of two functions are differentiated like

$$\frac{d}{dx}(P(x)\ y) = P(x)\frac{dy}{dx} + P'(x)\ y \tag{2.1.12}$$

where $P'(x)$ denotes the derivative of $P(x)$ with respect to x. Let us then multiply (2.1.11) by this unknown function $P(x)$ to obtain

$$P(x)\frac{dy}{dx} + P(x)\ p(x)\ y = P(x)\ q(x)$$

If we compare this with (2.1.12) we must have

$$P' = p(x)P$$

which is a separable differential equation as in (2.1.9). Hence, by (2.1.10)

$$\begin{aligned} \int \frac{1}{P}\ dP &= \int p(x)\ dx \\ \ln P &= \int p(x)\ dx \end{aligned}$$

$$P = e^{\int p(x)\ dx} \tag{2.1.13}$$

where the constant c was arbitrarily chosen to be zero, since we need only one function P. By (2.1.12) we now have

$$\begin{aligned} \frac{d}{dx}\left(y\ e^{\int p(x)\ dx}\right) &= e^{\int p(x)\ dx}\frac{dy}{dx} + p(x)\ y\ e^{\int p(x)\ dx} \\ &= q(x)\ e^{\int p(x)\ dx} \end{aligned}$$

To find y we need only integrate on both sides and divide by $P(x)$ afterwards.

$$y(x) = e^{-\int p(x)\,dx} \int_0^x q(t)\, e^{\int p(t)\,dt}\, dt + c\, e^{-\int p(x)\,dx} \tag{2.1.14}$$

Note that the complementary function $f(x)$ is an exponential function so that $f(0) \neq 0$, and hence the solution exists for any prescribed boundary value. The function P in (2.1.13) is known as the *integrating factor* of equations (2.1.4) and (2.1.11).

- Do : Exercise 35 in §2.12.

Homogeneous Equations

Definition 2.2

A function $f(x, y)$ is said to be homogeneous of degree n if the sum of the powers of x and y in each term of f is n. □

For example: $f(x, y) = x^3y - 4x^2y^2 + 3y^4$ is homogeneous of degree 4, but $g(x, y) = x^2y - 3x^3 + 5$ is not homogeneous due to the difference in degree between the first two terms and the last term.

Definition 2.3

If a first order differential equation can be written in the form

$$\frac{dy}{dx} = \frac{f(x, y)}{g(x, y)} \tag{2.1.15}$$

where f and g are homogeneous functions of the same degree, then the equation is said to be homogeneous. □

For example:

$$\frac{dy}{dx} = \frac{xy}{x^2 + y^2} \tag{2.1.16}$$

is homogeneous.

There is a standard method to solve a homogeneous equation by introducing a new dependent variable

$$v = \frac{y}{x} \tag{2.1.17}$$

Since $y = vx$ we have

$$\frac{dy}{dx} = v + x\frac{dv}{dx}$$

and since f and g are both homogeneous of the same degree n, we need only divide f and g by x^n to obtain an expression in v on the right hand side.

Hence

$$\begin{aligned} v + x\frac{dv}{dx} &= F(v) \\ \frac{dv}{dx} &= \frac{F(v) - v}{x} \end{aligned}$$

which is separable.

For example, the equation (2.1.16) becomes, after division by x^2 in the numerator and denominator on the right hand side

$$\begin{aligned} v + x\frac{dv}{dx} &= \frac{v}{1 + v^2} \\ x\frac{dv}{dx} &= \frac{-v^3}{1 + v^2} \\ \int \frac{1 + v^2}{v^3}\, dv &= -\int \frac{1}{x}\, dx + c \\ -\frac{1}{2v^2} + \ln|v| &= -\ln|x| + c \end{aligned}$$

Replace v by y, using (2.1.17), to obtain

$$\frac{x^2}{2y^2} = \ln|y| - c$$

If the boundary value $y(0) = \alpha$ is imposed, then

$$y^2 \ln\left|\frac{y}{\alpha}\right| = \frac{x^2}{2}$$

provided that $\alpha \neq 0$. The solution is in this case in implicit form, which means that the solution cannot be written in the form $y = f(x)$. However one could write the solution in the form $x = g(y)$, which is just as useful and informative.

- Do: Exercise 36 in §2.12.

2.2 Population Growth: Malthus Model

The problem in forecasting the change in a given population occurs frequently, and is, in fact, quite old. The population can consist of people, or fish, or cells in a tumor, or neutrons in a nuclear reactor. In general this problem is very difficult, with many complicated parameters in it. We shall now construct a very simple model along the lines of §1.1.

Let us state the problem clearly (*identification* in §1.1): On the basis of known data, predict the size of a population at some future date.

Let $N(t)$ denote the number of members of the population at time t. Then $N(t)$ is a non-negative integer with the result that the graph of $N(t)$ as a function of t must be a step function, as in Figure 2.2.1 where a jump in the graph denotes births or deaths in the population.

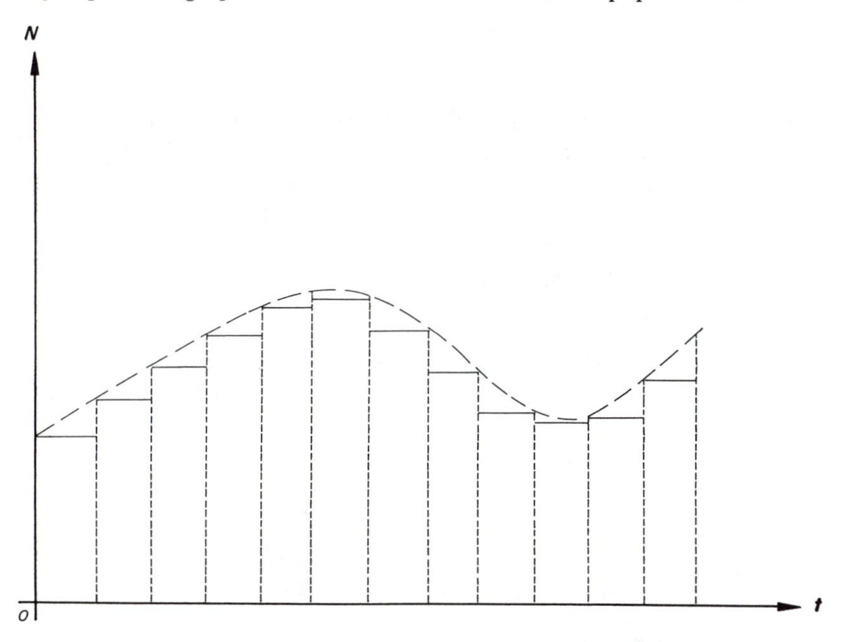

Figure 2.2.1: **Typical data of a population**

If $N(t)$ represents the population in a small town, this graph would be a realistic picture of the changes in the population. On the other hand, if $N(t)$ represents the population of China, a very short time would elapse between changes in the population. Hence, it would be reasonable to consider $N(t)$ as a continuous function of t. We shall make the following basic assumption:

Assumption (A)

The derivative of $N(t)$ is a continuous function of t for all $t > 0$.

The derivative of $N(t)$ describes the rate of change of the population. It is certainly true that more babies are born each day in China than in Iceland for the simple reason that there are many more people in China than in Iceland. So the change in population per day in China is much more than the change per day in Iceland; thus the rate of change of the population in China is bigger than the rate of change of the population in Iceland. This argument leads to the next assumption that the rate of change of the population is a function of the population:

$$\frac{dN}{dt} = f(N) \tag{2.2.1}$$

where f is a function which is known as the *growth function* of the population. One could also ask whether f is not possibly also dependent on the time t in the sense that the rate of change of the population may change from year to year. For example, the rate of change could be different in times of war (like the Second World War) than in times of peace and prosperity. We shall consider the simplified model where the growth function depends only on N as in (2.2.1).

Different models can now be constructed by assuming different growth functions in (2.2.1). For example, let us take the simplest case:

Assumption (B)

The rate of change of the population is directly proportional to the population at any time.

This completes the second stage in the modelling process (see Figure 1.1.1).

Assumptions (A) and (B) lead to the following initial value problem:

$$\frac{dN}{dt} = kN, \qquad N(0) = \alpha \qquad (0 < t < \infty) \tag{2.2.2}$$

where k is some constant and α denotes the initial population. The equations (2.2.2) represent the construction of the model (see Figure 1.1.1). Next we tackle the analysis of the model.

There are three ways in which $N(t)$ can be determined from (2.2.2).

Noting that for $k = 1$, the derivative of N, is equal to N, one can guess by the formula for the derivative of the exponential function that the solution must be of the form

$$N(t) = \alpha e^{kt} \tag{2.2.3}$$

Secondly, one can separate the variables (see §2.1)

$$\begin{aligned} \int \frac{1}{N}\, dN &= \int k\, dt + c \\ \ln N &= kt + c \end{aligned}$$

where c is a constant, to obtain (2.2.3).

Thirdly, an integrating factor can be used by multiplying equation (2.2.2) with e^{-kt} (see §2.1, especially (2.1.13)):

$$\begin{aligned} e^{-kt}\frac{dN}{dt} - ke^{-kt}N &= 0 \\ \frac{d}{dt}\left(Ne^{-kt}\right) &= 0 \\ Ne^{-kt} &= c \end{aligned}$$

(since N is a continuous function of t) which again gives (2.2.3). Hence a solution exists. We now also try to answer the other two questions posed in §1.2, namely uniqueness and sensitivity of the initial values.

Note that only the third method immediately shows uniqueness for all $t > 0$, since the first method was a guess and the second method is only valid if we know beforehand that $N(t) \neq 0$ for $t > 0$. In the third method, by the fundamental theorem of calculus (Theorem 2.1) all the solutions were found. For a given initial value α, (2.2.3) is then the only solution. We shall prove a general theorem on uniqueness in §2.14.

Is the solution a continuous function of the initial value? In this case it can easily be shown directly. Suppose $N(t)$ is the solution of

$$\frac{dM}{dt} = kM, \qquad M(0) = \alpha + \epsilon \tag{2.2.4}$$

where ϵ is a small real number. We know that the solution of (2.2.4) is

$$M(t) = (\alpha + \epsilon)e^{kt}. \tag{2.2.5}$$

Hence it follows from (2.2.3) and (2.2.5) that

$$\begin{aligned} |N(t) - M(t)| &= |\epsilon\, e^{kt}| \\ &\leq |\epsilon| e^{kT} \end{aligned}$$

if $0 \leq t \leq T$. The difference $|N(t) - M(t)|$ can be made arbitrarily small by choosing

$$|N(0) - M(0)| = |\epsilon|$$

small enough, for any t in the interval $[0, T]$.

So we are satisfied that the problem (2.2.2) has the unique solution (2.2.3), and now we can interpret this solution of the mathematical model of population growth. If k is positive, the population will grow exponentially. If k is zero, the population will remain constant at α. If k is negative, the population will diminish, but will never be zero, mathematically speaking. In actual fact, of course, if $N(t)$ is less than one, then no member of the population is left. The value of k is thus of critical importance for the behaviour of the solution. To determine k, the size of the population at some time instant $t_1 \neq 0$ must be known. Once k is known, $N(t)$ is known at any time t, and we can calculate the population at a given time t, or the time necessary for the population to attain a prescribed size.

Note that k is a growth rate per population with the dimension of $(\text{time})^{-1}$. In some situations k is known. It can be expressed as "30 per thousand per year" or equivalently as 3% or as 0,03. Actually k is the net growth rate per population. We can also assume that k is composed of two parts namely $k = b - d$ where b represents the constant growth rate due to births and d represents the constant decay rate due to deaths.

- Do : Exercises 1, 2, 3 in §2.12.

The model (2.2.2) was formulated by the English demographer and economist, Thomas R. Malthus (1766-1834) in "An Essay on the Principle of Population" in 1798. At the time of the industrial revolution the population of Europe was rapidly increasing. Malthus showed that the production of food increased linearly, while the population increased exponentially, according to his model. On this basis he predicted a crisis in the future. However, the crisis never materialized, probably because of massive emigration to America and Australia, as well as the wars in Europe during the nineteenth century. This drain on the population was not built into the model of Malthus, so that his model could not predict the change in population correctly. We shall look into this deficiency of the model in §2.4.

Finally, the model must be validated. We shall discuss this very important aspect in the next section.

2.3 Least Squares Method of Curve Fitting

Suppose the number of cells in a tumor is counted every day so that the values in Figure 2.3.1 are obtained where $N(t)$ is measured in thousands.

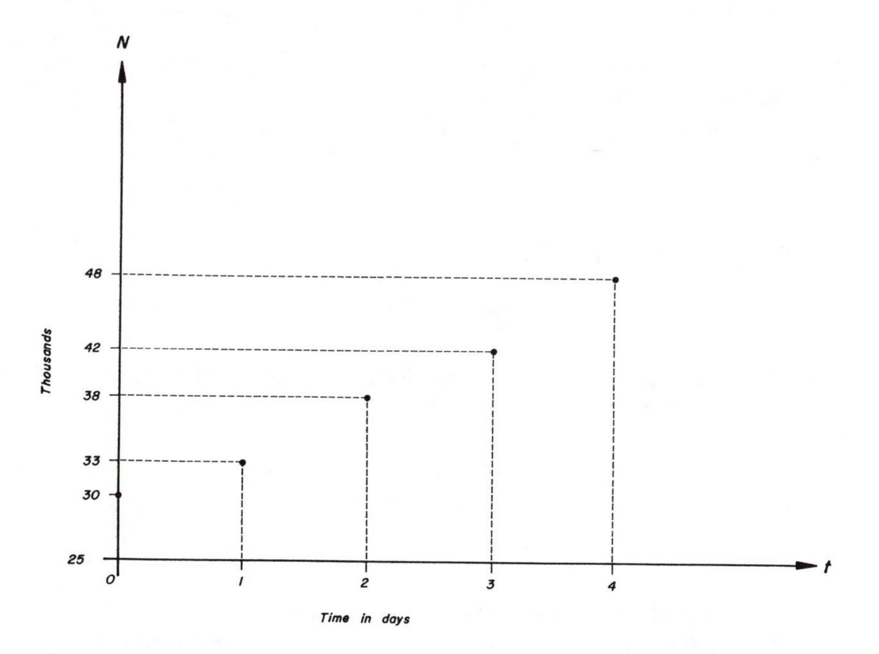

Figure 2.3.1: **Population of a tumor over a period of five years**

It is easy to find several curves which all pass through the given five points: for example, the fourth degree polynomial

$$\begin{aligned} N(t) &= 30\frac{(t-1)(t-2)(t-3)(t-4)}{(0-1)(0-2)(0-3)(0-4)} + 33\frac{t(t-2)(t-3)(t-4)}{(1)(1-2)(1-3)(1-4)} \\ &\quad +38\frac{t(t-1)(t-3)(t-4)}{2(2-1)(2-3)(3-4)} + 42\frac{t(t-1)(t-2)(t-4)}{3(3-1)(3-2)(3-4)} \\ &\quad +48\frac{t(t-1)(t-2)(t-3)}{4(4-1)(4-2)(4-3)} \\ &= \frac{t^4}{4} - 2t^3 + \frac{21t^2}{4} - \frac{t}{2} + 30 \end{aligned}$$

In fact, there are infinitely many fifth degree polynomials passing through the given five points, by prescribing an arbitrary value at $t = 5$.

The essential point is that the interpolation between a given discrete set of data is arbitrary, unless some qualitative information is available or unless some reasonable assumptions can be made on the growth rate of the tumor. Suppose we decide that Assumption (B) is reasonable in this case. Then we have the same boundary value problem (2.2.2) with the unique solution (2.2.3) as before:

$$N(t) = \alpha e^{kt}$$

The initial value $\alpha = 30$ is given, and k can be calculated by using the fact that $N(1) = 33$. We find that

$$k = \ln\left(\frac{33}{30}\right) = 0.0953$$

so that the equation for $N(t)$ becomes

$$N(t) = 30e^{0.0953t}$$

The other data points can now be calculated to check the accuracy of the model:

$$\begin{aligned} N(2) &= 36.3 \\ N(3) &= 39.93 \\ N(4) &= 43.923 \end{aligned}$$

Unfortunately, this does not agree with the values in Figure 2.2.1. There are two possible explanations for this: either Assumption (B) was wrong, or there are some experimental errors in the data.

Let us try a clever trick to gain insight into our dilemma. Take logarithms on both sides of (2.2.3):

$$\ln N(t) = \ln \alpha + kt \tag{2.3.1}$$

If we draw the graph of (2.3.1) with $\ln N(t)$ on the vertical axis and t on the horizontal axis, then (2.3.1) represents a straight line with slope k which intersects the vertical axis at $\ln \alpha$. The experimental values $E(i)$ of Figure 2.3.1 (with $i = 0, 1, 2, 3, 4$) can also be plotted as the points $(i, \ln E(i))$, as shown in Figure 2.3.2. Since the graph of (2.3.1) is a straight line, we can try to shift a ruler on Figure 2.3.2 to fit the five data points. It is immediately clear that there are many approximate answers, but no single straight line which passes through all five points. So we are confronted with the question: which line (or which k and α) gives the best fit to the given data points?

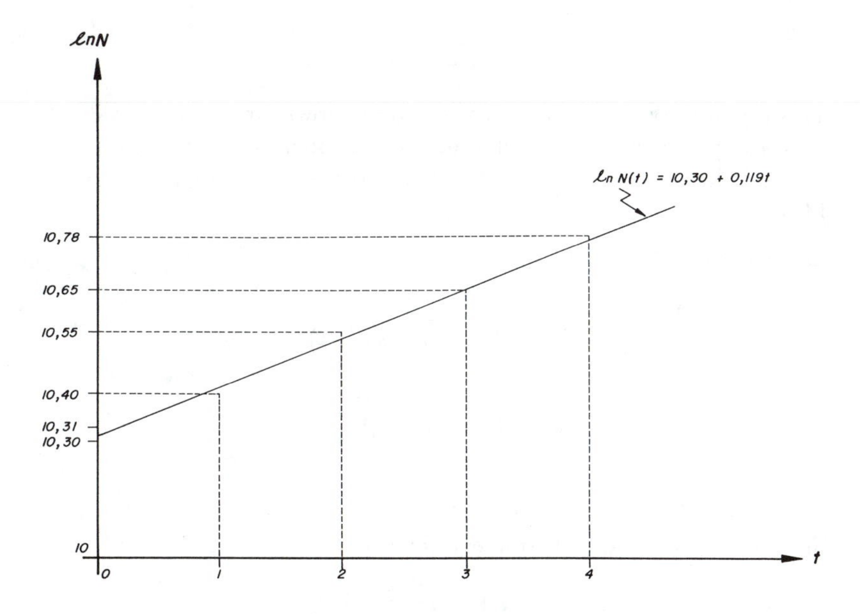

Figure 2.3.2: **Logarithms of the data in Figure 2.3.1**

A popular answer to this question is the method of least squares. It means simply that the sum of the squares of the distances between the experimental points and the values of the graph must be as small as possible. (Do you remember Exercise 1 in §1.5?) In mathematical terms we can say:

Minimize the expression

$$\sum_{i=0}^{4}[\ln E(i) - \ln N(i)]^2$$

or, if we utilize (2.3.1):

Choose k and α such that

$$F(\ln \alpha, k) = \sum_{i=0}^{4}[\ln E(i) - \ln \alpha - ki]^2 \tag{2.3.2}$$

is a minimum. The function F depends on the two independent variables k and $\ln \alpha$. To find the minimum, we differentiate F with respect to k and $\ln \alpha$ and equate the two partial derivatives to zero:

$$\begin{aligned} \frac{\partial F}{\partial k} &= 0 \\ \frac{\partial F}{\partial (\ln \alpha)} &= 0 \end{aligned}$$

This provides two equations in the two unknowns $\ln \alpha$ and k. It can be shown that it is always possible to solve for $\ln \alpha$ and k, and that these values of $\ln \alpha$ and k will in fact give the required minimum (see [64] p. 130).

In the case of (2.3.2) we find

$$\begin{aligned} \sum_{i=0}^{4} 2[\ln E(i) - \ln \alpha - ki](-i) &= 0 \\ \sum_{i=0}^{4} 2[\ln E(i) - \ln \alpha - ki](-1) &= 0 \end{aligned}$$

After the substitution of the values of $E(i)$ and some simplification, these equations lead to two linear equations:

$$\begin{aligned} 5 \ln \alpha + 10k &= 52.69 \\ 10 \ln \alpha + 30k &= 105.57 \end{aligned}$$

with the solution

$$k = 0.119 \qquad \text{and} \qquad \ln \alpha = 10.30$$

The line $\ln N(t) = 10.30 + 0.119t$ is shown in Figure 2.3.2. The error is small, and hence we conclude that Assumption (B) was reasonable in this case. Finally then, we predict the growth of the tumor by the formula

$$N(t) = 19.73e^{0.119t}$$

where the cells are measured in thousands.

- Read: [61] pages 119 to 138.
- Do: Exercises 4, 5 in §2.12.
- Do: Project A* in §2.13.

2.4 Population Growth: Logistic Model

In §2.2 and §2.3 we went through all the stages of mathematical modelling (see Figure 1.1.1) and produced a model which can be used to predict the growth of a tumor. In this section we consider a case when the validation is unsatisfactory. For this purpose we look at the population of the United States of America.

The census figures of population of the United States of America are shown in Figure 2.4.1, where the vertical axis represents the population N in millions and the horizontal axis represents the time t in years with $t = 0$ chosen in the year 1790.

As in §2.2 we aim to construct a simple model which can predict the population growth in the United States of America. If we try the Malthus model, and plot $\ln N$ in Figure 2.4.2, we immediately see that the data points do not lie on a straight line. It seems rather like two straight lines: the first one through the points at $t = 0,\ 10,$ and 60 with a steeper slope k than the one through the other points. So we have to devise a new model in which k is not a constant, but changes with time. Hence, in the diagram of Figure 1.1.1 we must return to the assumptions stage to rethink the underlying mechanism as it was embodied in Assumption (B).

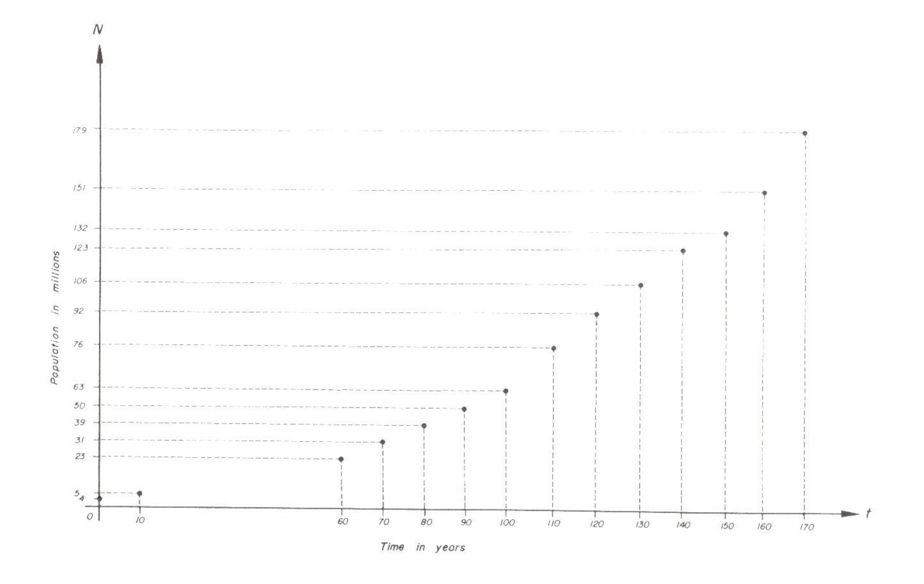

Figure: 2.4.1: **Population of the United States of America in the period 1790 - 1960**

As we had seen in §2.2, the constant k in (2.2.2) represents a net growth rate per population, which is composed of two parts namely $k = b - d$ where b represents the growth rate due to births and d the decay rate due to deaths. Since k in Figure 2.4.2 seems to be diminishing, we may assume that d is an increasing function of N. This would, for example, be the case in a population where the food supply is limited. Let us make the simplest assumption and see whether the resulting model would fit the data in Figure 2.4.1:

Assumption (C)

The rate of change of the population N is of the form $(b - sN)N$ where b and s are positive constants.

Assumption (C) followed from a physical argument based on the population figures. We could also use a mathematical argument in the sense that Assumption (B) on the growth function $f(N)$ of (2.2.1) must be replaced by some other function. Let us suppose that $f(N)$ can be written as a Maclaurin series (see §2.13):

$$f(N) = a_0 + a_1 N + a_2 N^2 + \ldots$$

Obviously $f(0) = 0$ because if there are no people, the growth rate must be zero. Hence $a_0 = 0$. If we approximate $f(N)$ by the second term, we obtain the Malthus model. The next logical step is to approximate $f(N)$ by $a_1 N + a_2 N^2$ which is precisely Assumption (C). (This argument was used by Lotka in [40].)

We also note in Figure 2.4.1 that the slope of $N(t)$ is positive at $t = 0$. This fact is stated in the next assumption:

Assumption (D)

If α denotes the population at $t = 0$, then $b - s\alpha > 0$.

The mathematical model resulting from Assumptions (A), (C), and (D), for the population growth in the United States of America is the following initial value problem:

$$\frac{dN}{dt} = (b - sN)N, \qquad N(0) = \alpha \qquad (0 < t < \infty) \tag{2.4.1}$$

Note that the differential equation is nonlinear due to the term sN^2, and also separable (see §2.1). By Assumption (D) it follows that $b - sN(t) > 0$ and $N(t) > 0$ for all t in some interval $0 \le t \le T$, since $N(t)$ is a continuous function (see the definition of a solution in §1.2). Hence for $0 \le t \le T$ we can rewrite (2.4.1) as

$$\int_\alpha^N \frac{dx}{(b - sx)x} = t \tag{2.4.2}$$

Use partial fractions in the integral (or see Table of Integrals: Chapter 8, line 1) to obtain

$$\ln\left(\frac{N}{b - sN}\right) - \ln\left(\frac{\alpha}{b - s\alpha}\right) = bt$$

$$\frac{N}{b - sN} = \frac{\alpha}{b - s\alpha} e^{bt} \tag{2.4.3}$$

$$N(t) = \frac{b\alpha}{s\alpha + (b - s\alpha)e^{-bt}} \tag{2.4.4}$$

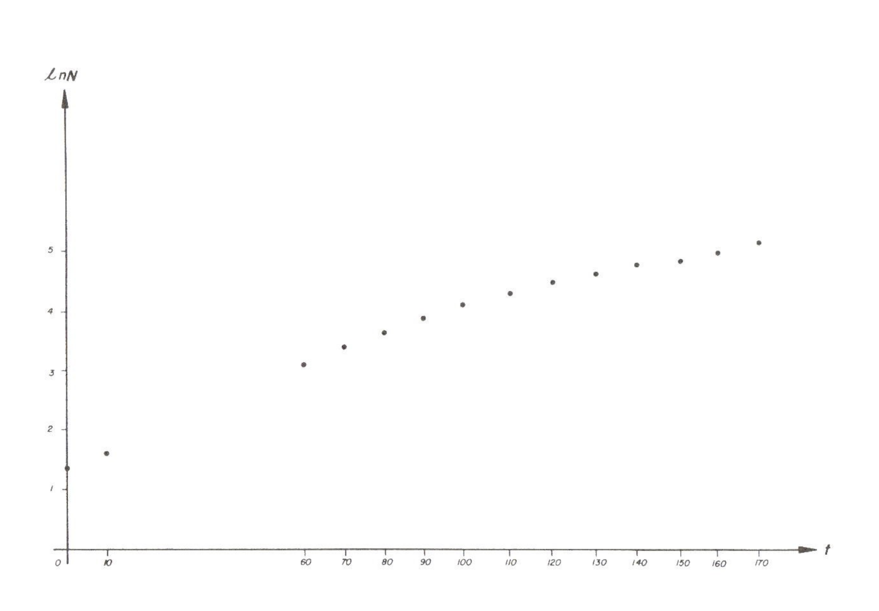

Figure 2.4.2: **Logarithms of the data in Figure 2.4.1**

By direct differentiation it can be verified that (2.4.4) is a solution for the initial value problem (2.4.1) for all $t \geq 0$. Hence the solution exists. To investigate the uniqueness of this solution, we shall use a general theorem, but first we need an important definition.

Definition 2.4

A function $f(x)$ satisfies a Lipschitz condition *on the interval I if there exists a constant k such that*

$$|f(r) - f(s)| \leq k|r - s| \tag{2.4.5}$$

for all r and s in I. □

The word *Lipschitz* in the definition refers to Rudolf Lipschitz who first introduced this condition in 1876.

If a function satisfies a Lipschitz condition on an interval I, then obviously it must be continuous there, but the converse is not true (see

Exercise 7 in §2.12). On the other hand, if f has a continuous derivative on a closed interval I, then f automatically satisfies a Lipschitz condition. This follows from the mean-value theorem ([16] p. 174):

$$f(r) - f(s) = (r - s)f'(\xi), \qquad r < \xi < s$$

Since the derivative f' is continuous on a closed interval, it must be bounded on I, so that (2.4.5) is obtained by taking absolute values on both sides. Note also that the function $f(x) = |x|$ satisfies a Lipschitz condition on any interval, although the derivative of f does not exist at $x = 0$.

- Do: Exercises 6, 7, 8 in §2.12.

We can now state a general theorem on uniqueness (see §2.14 for the proof):

Theorem 2.4.1

Let $f(t, y)$ be continuous on the rectangle $R = \{(t, y) : a \leq t \leq b,\ c \leq y \leq d\}$ and let $f(t, y)$ satisfy a Lipschitz condition in y on $[c, d]$ with the same constant k for every t in $[a, b]$. If a solution of the initial value problem

$$\frac{dy}{dt} = f(t, y), \qquad y(a) = \alpha \tag{2.4.6}$$

exists in R, then the solution is unique and the solution is a continuous function of the boundary value α in R. □

Let us apply this theorem to the model (2.4.1) of the populaton of the United States of America. In this case $f(t, N) = bN - sN^2$, which has a continuous derivative $b - 2sN$ with respect to N. Hence, f satisfies a Lipschitz condition in N on any finite interval, and since f is not a function of t, the same constant k will hold for all $t \geq 0$. The theorem now states that as long as the solution N is bounded, so that d is finite, $N(t)$ is the unique solution of (2.4.1). But from (2.4.4) it is clear that $N(t)$ has a limit $\frac{b}{s}$ as $t \longrightarrow \infty$ and it is thus bounded. Therefore, this solution is the only solution for $t \geq 0$. Note that the continuous dependence on the boundary value α can also be shown directly from (2.4.4).

In 1920 R. Pearl and L. J. Read applied this model to the population of the United States of America. They used the data 1790, 1850, and 1910 in Figure 2.4.1 (that is $t = 0$, 60, and 120) to obtain by (2.4.4) three equations in the three unknowns a, b, and s. To find the solution of this system of three equations is not easy and leads to a nonlinear equation which cannot be solved explicitly. (See Project B^* in §2.13 for more details.) However, once a, b, and s are found, they can be substituted in (2.4.4) to give

$$N(t) = \frac{210}{1 + 51.5e^{-0.03t}}$$

(You can check the accuracy of the result by substituting $t = 0$, 60, and 120 - you should find that $N(0) = 4$, $N(60) = 23$, and $N(120) = 92$.) The graph of this function is shown in Figure 2.4.3 with the data of Figure 2.4.1. The comparison of the predicted values of the model with the actual population in the years after 1920, is shown in the following table:

Year	t	N(t)	Population
1920	130	103.5	106
1930	140	118.5	123
1940	150	133.6	132
1950	160	147.4	151
1960	170	159.8	179
1970	180	170.4	205
1980	190	179.1	226

The prediction is fairly accurate up to 1950 with an error of less than 3%, but the figure for 1960 shows clearly that the model is not reliable for a longer period of time. In fact, according to the model the population of the United States of America should stabilize at 210 million, while it has already exceeded that number in 1980, and was 241.6 million in 1986. Another factor to take into account is the accuracy of the census figures given in Figure 2.4.1, since these values were used to determine the parameters a, b, and s of the model.

The assumption of a constant birth rate b in the model is not actually true. In the period 1910 to 1960 the birth rate fluctuated between 18.7 per thousand per year and 30 per thousand per year. One would also expect that the Civil War and the two world wars must have influenced the values of b and d in the years concerned. Moreover, in the period 1800 to 1950 about 40 million people from Europe emigrated to the United States of America - a fact which was ignored in the construction of the model (see Exercise 2 in §2.12). It is, therefore, rather surprising that the model did give fairly accurate results up to 1950.

The Equation (2.4.1) is known as the *logistic equation*, and the corresponding graph in Figure 2.4.3 as the logistic curve. (Some books call it the S-curve.) The Belgian sociologist P. F. Verhulst used this equation in 1847 to study the human population, and since then it has been used in many instances for both theoretical and experimental research. (Read, for example, the work by A. J. Lotka in [40].)

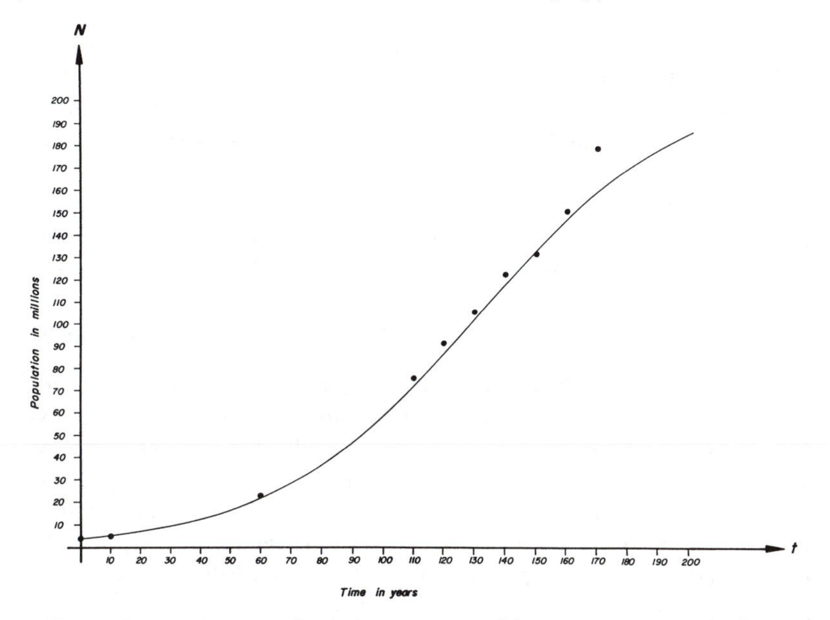

Figure 2.4.3: **Comparison between $N(t)$ and the population data**

The logistic equation is often used by biologists. Apart from Assumptions (A), (B), and (D) there are also other simplifying assumptions built into it. According to E. C. Pielou (see [49] p. 35) these assumptions are:

(1) Abiotic environment factors are sufficiently constant not to affect the birth and death rates.

(2) Crowding affects all population members equally. (This is unlikely to be true if the individuals occur in clumps instead of being evenly distributed throughout the available space.)

(3) Birth and death rates respond instantly, without lag, to density changes.

(4) Population growth rate is density-dependent, even at the lowest densities. (It may be more reasonable to suppose that there is some threshold density below which individuals do not interfere with one another.)

(5) The population has, and maintains, a stable age distribution.

(6) The females in a sexually reproducing population always find mates, even when the density is low.

If anyone of these assumptions is false, then a modified model must be constructed.

In 1963 F. E. Smith made a study of the bacteria *Daphnia magna* and found that neither the Malthus nor the logistic models could fit the data satisfactorily. In [55] he constructed a new model, arguing as follows: If food is the limiting factor in the growth of the population, then the growth rate kN should be multiplied by a fraction representing in the numerator the rate of food supply not momentarily being used by the population and in the denominator the rate of food supply of the maximal population size N_0. Let us write instead of (2.4.1)

$$\frac{dN}{dt} = kN\frac{S-F}{S} \tag{2.4.7}$$

where F is the rate at which a population of size N uses food and S is the corresponding rate when the population reaches saturation level. When the population increases, food is used for both maintenance and growth; but at saturation level, food is used for maintenance only, since growth had stopped. Hence, F must be a function of N (the size of the population being maintained) and $\frac{dN}{dt}$ (the rate at which the population is growing). Let us consider the simplest case, namely

$$F = pN + q\frac{dN}{dt} \tag{2.4.8}$$

where p and q are positive constants. When the saturation level N_0 of the population is reached, we must have $\frac{dN}{dt} = 0$ and hence, by (2.4.8) it follows that $S = pN_0$. Substitute these values of F and S in (2.4.7) and simplify to obtain

$$\frac{dN}{dt} = kN\frac{b-N}{b+cN} \tag{2.4.9}$$

where $b = N_0$ and $c = kqp^{-1}$ are positive constants. Note that $c = 0$ (or equivalently $q = 0$) implies the logistic equation. With the model (2.4.9) Smith obtained a reasonable fit to his experimental data.

Many better and more complicated models for population growth have been formulated. One example can be found in [26] where different age groups in the population are taken into account in the model. Very young and very old people make no contribution to the birth rate and the latter group is the major influence on the mortality rate. This provides a more realistic model, but also a much more complicated one.

- Read: [50] pages 83 to 85, [53] pages 185 and 186 (where the same problem is formulated as an integral equation), and [12].
- Do: Exercises 9, 10, 11 in §2.12.
- Do: Projects B^* and C in §2.13.

2.5 Harvesting

Sometimes a mathematical model only requires a qualitative description of the solution of the differential equation concerned, in which case it is not even necessary to solve the differential equation! In this section we discuss such a model.

The harvesting of renewable resources is studied intensively because man is inclined to ruin the resource in his greed, instead of limiting himself to an optimal harvest. Many fishing grounds collapsed under overfishing and some animals are in danger of extinction due to indiscriminate hunting. Let us consider a population of animals of which an annual harvest is taken. Suppose the logistic model fits the normal growth of the population (without harvesting) and suppose that a constant number E of the population are killed every year. If N represents the population at time t, then the initial value problem (2.4.1) changes to

$$\frac{dN}{dt} = (b - sN)N - E, \qquad N(0) = \alpha \qquad (t > 0) \qquad (2.5.1)$$

If the constants b, s, α, and E are known, the solution $N(t)$ can be determined. However, in this case we are not so much interested in the value of N at a specific time instant t, but rather in the terminal value of N as $t \longrightarrow \infty$. The crucial question is whether the population will die out in a finite period of time. If not, will N tend to a limit β as $t \longrightarrow \infty$, as was the case in §2.4 with $E = 0$? Note that if (2.5.1) is compared to (2.4.1), then the slope of the graph of $N(t)$ in (2.5.1) is less than the slope of the graph of $N(t)$ in (2.4.1) at any time t for the same N and the same parameters b and s. Since the graph of (2.4.1) tends to a limit as $t \longrightarrow \infty$, one would expect the same behaviour for the graph of (2.5.1).

The terminal value of N is important to ecologists who guard against the extinction of animal or botanical species, as well as to scientists in agriculture who have to control pests. Especially in the calculation of fishing quotas, scientists try to choose E in such a way that the annual catch is as large as possible without diminishing the stock (the maximum sustainable yield). Thus the problem is to determine the limit β as a function of E.

At the limit β the slope of $N(t)$ is zero, and so (2.5.1) reads

$$\begin{aligned} 0 &= (b - s\beta)\beta - E \\ s\beta^2 - b\beta + E &= 0 \\ \beta &= \frac{b \pm \sqrt{b^2 - 4sE}}{2s} \end{aligned}$$

If $b^2 - 4sE \geq 0$ we obtain real values for β. Let

$$F = \frac{b^2}{4s} \tag{2.5.2}$$

then $E \leq F$ will imply the existence of a limit. On the other hand $E > F$ means that

$$(b - sN)N - E < 0$$

so that the graph of N has a negative slope and hence, the population will be extinct in a finite time T, as shown in Figure 2.5.1. Therefore, F is a critical value in the sense that a harvesting rate which exceeds F must lead to the collapse of the stock.

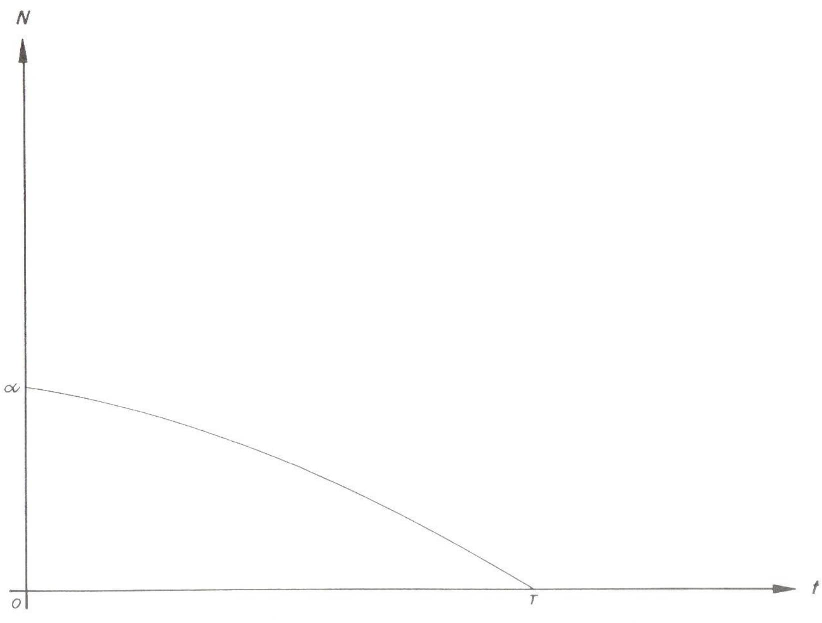

Figure 2.5.1: **Qualitative behaviour of $N(t)$ when $E > F$**

We can take the qualitative analysis even further. Let $E < F$. Then there are two different limiting values β_1 and β_2 where

$$\beta_1 = \frac{b}{2s} + \frac{\sqrt{b^2 - 4sE}}{2s} = \frac{\sqrt{F} + \sqrt{F - E}}{\sqrt{s}}$$

$$\beta_2 = \frac{b}{2s} - \frac{\sqrt{b^2 - 4sE}}{2s} = \frac{\sqrt{F} - \sqrt{F - E}}{\sqrt{s}}$$

Equation (2.5.1) can be rewritten as

$$\frac{dN}{dt} = -s(N - \beta_1)(N - \beta_2) \tag{2.5.3}$$

Note that the solution of (2.5.1) is unique by Theorem 2.4.1 (as in the case of (2.4.1)). *This means that for different initial values the graphs of the corresponding solutions may never intersect.* For if the graphs do intersect at $t = \lambda$, then consider the differential equation on the interval $\lambda \le t < \infty$ with initial value $N(\lambda)$. Since there are two solutions, both with initial value $N(\lambda)$, the uniqueness guaranteed by Theorem 2.4.1 is contradicted.

We can now draw a qualitative graph of (2.5.3) as shown in Figure 2.5.2. If $\alpha > \beta_1$ (and thus also $\alpha > \beta_2$) then the slope at $t = 0$ is negative, and the slope will remain negative as long as $N(t) > \beta_1$. But also note that if $\alpha = \beta_1$, then the slope is zero and hence, $N(t) = \beta_1$ remains constant. By the above argument the graph for $N(t)$ for the case $\alpha > \beta_1$ may not intersect $N(t) = \beta_1$, although its slope is strictly negative. Hence the slope must tend to zero so that $N = \beta_1$ is always the limit when $\alpha > \beta_1$.

If $\beta_2 < \alpha < \beta_1$ the slope at $t = 0$ is positive, and remains positive as long as $N < \beta_1$. Since the graph may not intersect the line $N = \beta_1$, it follows by the same argument as above that $N = \beta_1$ is always the limit in this case.

If $\alpha = \beta_2$ then again $N(t) = \beta_2$ is the constant solution to (2.5.3).

Finally, if $0 < \alpha < \beta_2$ then the slope at $t = 0$ is negative and the slope becomes steeper as N gets smaller, so that N must be zero within a finite time T. We call T the *extinction time*. So even if $E < F$, then population can still become extinct if the initial population is below the critical value of β_2.

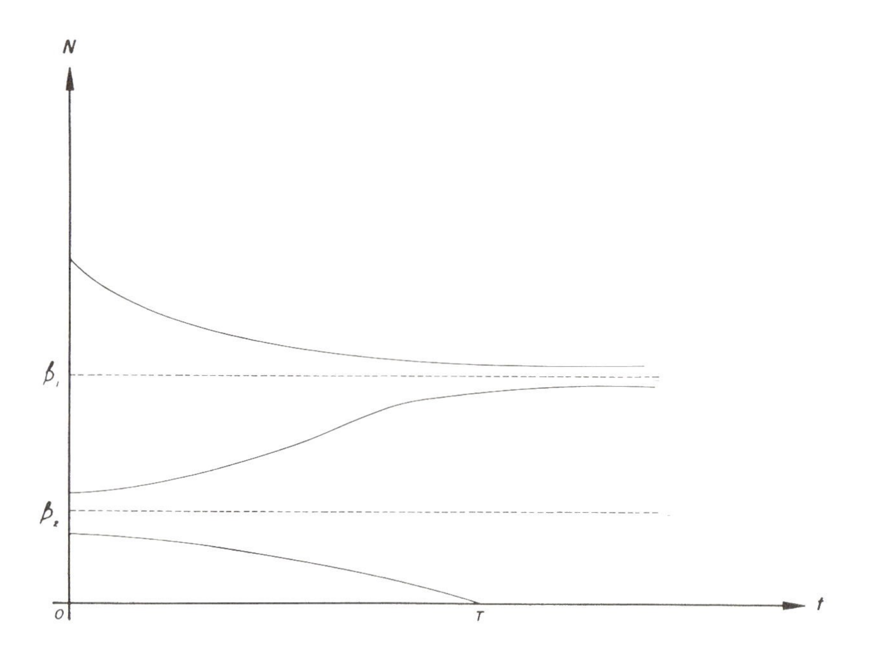

Figure 2.5.2: **Qualitative behaviour of $N(t)$ when $E < F$**

The extinction time T can be obtained by integrating (2.5.1):

$$T = \int_{\alpha}^{0} \frac{dN}{(b - sN)N - E} \tag{2.5.4}$$

The case when $E = F$ is simpler because there is only one value for the limit $\beta_1 = \beta_2 = \beta$. If $\alpha > \beta$ then N will tend to β as $t \longrightarrow \infty$, and if $\alpha < \beta$ then N will be zero in a finite time T.

We now have insight in the qualitative behaviour of the solution of (2.5.1). Let us use the knowledge in the following model:

The sandhill crane *Grus canadensis* in North America was protected since 1916 because it was on the endangered list. Repeated complaints of crop damage in the Western United States and Canada led to hunting seasons since 1961. These birds will not breed until they are 4 years of age and normally will have a maximum life span of 25 years. Two American ecologists, R. Miller and D. Botkin published a report [44] in which they constructed a simulation model with ten parameters to investigate the effect of different rates of hunting on the sandhill crane. If one fits our simple model (2.5.1) to their data to compare our model with their more sophisticated model, one finds that

$$b = 0.09866 \qquad \text{and} \qquad s = 5.06989 \times 10^{-7} \tag{2.5.5}$$

By (2.5.2) it follows that the critical hunting rate is $F = 4800$ birds per year. Take the initial value $\alpha = 194,600$ which is the limit bs^{-1} of the logistic model when $E = 0$ (see (2.4.4)). The following extinction times can then be calculated from (2.5.4):

E (in thousands)	T (in years)	Miller and Botkin
6	90	71
8	44	38
12.77	21	19

If values of E less than F are taken and the limit β_1 calculated, we find:

E (in thousands)	β (in thousands)	Miller and Botkin
2	171.6	167.5
3	156.9	149.9
4	137.0	125.8

Our model is more optimistic than the prediction of Miller and Botkin, and surprisingly near to it, considering the simplicity of our assumptions. Finally, Miller and Botkin have bad news for the sandhill cranes. They estimate in [44] that 9800 sandhill cranes were legally bagged in 1970, and that 30% more were killed by illegal hunters or died later of wounds - a total of 12,770 or 6% of the population. As was shown above, this means that the sandhill crane may become extinct within a period of 19 years. Miller and Botkin plead for stricter control on indiscriminate hunting and for smaller quotas.

- Read: [11] pages 13 to 16. A more general model for the harvesting of tuna is in [48]. The story of the collapse of the anchovy stock on the coast of Chile in [34] illuminates the complexity of these problems.

- Do: Exercises 12, 13, 14, 15 in §2.12.

- Do: Project D^* in §2.13.

2.6 Optimization of Profit

Many problems in economics lead to mathematical models in which differential equations play a prominent role. We shall consider a simple typical example. Note that in §2.4 the solution of the differential equation itself was important, but in §2.5 the qualitative behaviour of the solution was required (in fact we never solved the initial value problem explicitly). In this case study, the solutions of the differential equations are going to play a minor role with a lot of hard work after the solutions had been obtained.

The first step in the modelling process is the identification (see Figure 1.1.1). Let us consider an idealized company. The object of the management is to produce the best possible dividend for the shareholders. We assume that the bigger the capital invested in the company, the bigger will be the profit (the net income). This means that if the total profit is paid out as a dividend to the shareholders, then the company does not grow. In the long term the management would like to re-invest a part of the profit annually in the company so that the subsequent profits in future years will increase. The problem is what part of the profit must be paid out annually as a dividend so that the total yield for the shareholder over a given period of years is a maximum.

Next we consider the assumptions necessary to construct a mathematical model. Let $u(t)$ represent the capital invested in the company at time t. Normally the capital and profit will be calculated at the end of the financial year of the company, so that the graph of $u(t)$ would be a step function as in Figure 2.2.1. We shall, however, make the assumption that the capital and profit are known continuously and that the process of re-investment and dividends is also continuous. Then $u(t)$ is a continuous function and we assume:

Assumption (E)

The function $u(t)$ is a differentiable function of t for $0 < t < \infty$.

We also need an assumption on the amount of profit produced in a given time interval.

Assumption (F)

The profit in the interval $[t,\ t+\delta t]$ is directly proportional to $u(t)\delta t$ for an arbitrarily small δt.

We can now move on to the next stage of the modelling process, namely the construction of the model. Denote the total dividend in the period $[0,t]$ to the shareholders by $w(t)$. Let k be the constant fraction of the profit produced in $[t,\ t+\delta t]$ which will be re-invested as the addition δu to the capital for this period. Then $0 \leq k \leq 1$. It follows that

$$\delta u = kau(t)\delta t$$

where a is the proportionality constant in Assumption (F). Similarly if δw represents the dividend paid out of the rest of the profit produced in $[t,\ t+\delta t]$, we must have

$$\delta w = (1-k)au(t)\delta t$$

Divide both equations by δt and let $\delta t \longrightarrow 0$, then we obtain two differential equations

$$\frac{du}{dt} = kau, \qquad u(0) = \alpha \tag{2.6.1}$$

$$\frac{dw}{dt} = a(1-k)u, \qquad w(0) = 0 \tag{2.6.2}$$

with the initial capital denoted by α and the initial dividend is, of course, zero. The analysis of the model (2.6.1) and (2.6.2) is simple. We recognize (2.6.1) as a Malthus model with the solution

$$u(t) = \alpha e^{akt}, \tag{2.6.3}$$

Substitute (2.6.3) in (2.6.2) and integrate to obtain

$$w(t) = \begin{cases} \frac{(1-k)\alpha}{k}(e^{akt}-1) & \text{for } 0 < k \leq 1 \\ a\alpha t & \text{for } k = 0 \end{cases} \tag{2.6.4}$$

Note that if $k = 1$, we have $w(t) = 0$ for $t \geq 0$; in other words, no dividend is paid out to the shareholders, but the capital of the company grows instead. On the other hand if $k = 0$, the capital remains constant at α and all the profit is paid out as a dividend. The central issue here is:

> *Given a period of time $[0, T]$, how must k be chosen so that the total dividend over the period $[0, T]$ is a maximum?*

By (2.6.4) we have for $0 < k \leq 1$ that

$$w(T) = \frac{(1-k)\alpha}{k}\left(e^{akT} - 1\right) \tag{2.6.5}$$

To simplify the calculations, let us use scaling. Introduce new variables

$$x = aTk \qquad \text{and} \qquad y = \frac{w(T)}{\alpha} \tag{2.6.6}$$

Then, since $0 < k \leq 1$, we must have $0 < x \leq aT$ and (2.6.5) transforms to

$$y = \frac{aT - x}{x}(e^x - 1) \tag{2.6.7}$$

Since a, T, and α are known constants, the question still is:

> *For which values of x will y be a maximum?*

The obvious way to answer the question is to differentiate y with respect to x and to determine for which values of x the derivative will be zero. Hopefully this will produce the required maximum of y.

$$\left.\begin{aligned}\frac{dy}{dx} &= aTx^{-2}e^x\left[-1 + x - (aT)^{-1}x^2 + e^{-x}\right] \\ &= aTe^x\left[-\tfrac{1}{aT} + \tfrac{1}{2} - \tfrac{x}{3!} + \tfrac{x^2}{4!} - \tfrac{x^3}{5!} + \ldots\right]\end{aligned}\right\} \tag{2.6.8}$$

where the Maclaurin series of e^{-x} was used (see §2.13). Let us consider three cases:

(1) If $aT = 2$, the first two terms of the series in (2.6.8) cancel out, and since $0 < x \leq 2$, the remaining series is alternating and monotonic decreasing. Hence, the sum of the series is negative (see (2.14.7) in §2.14) and the derivative is never zero for $0 \leq x \leq 2$. Thus y is a decreasing function of x with the maximum of y at $x = 0$.

(2) If $aT < 2$, the sum of the first two terms of the series in (2.6.8) is negative, and by the same argument y is again a decreasing function with the maximum of y at $x = 0$.

(3) If $aT > 2$, the slope of y is positive at $x = 0$ by (2.6.8). However, it also follows from the first line of (2.6.8) that the slope of y is negative at $x = aT$. Since the slope of y is a continuous function of x on the closed interval $[0, aT]$, there is at least one point $X \in (0, aT)$ where the slope is zero. By (2.6.8) X must satisfy the equation

$$1 - x + \frac{x^2}{aT} = e^{-x} \tag{2.6.9}$$

The left hand side is a quadratic function of x whose graph is a parabola symmetric to the line $x = \frac{aT}{2}$ and intersecting the vertical z−axis at the point $z = 1$. The right hand side is an exponential curve with negative slope and also intersecting the vertical z−axis at the point $z = 1$. These curves are shown in Figure 2.6.1. It is clear that there can be, at most, one point of intersection in the open interval $(0, aT)$. Hence, the point $X \in (0, aT)$ is unique. Since the slope of y is positive for $0 < x < X$ and negative for $X < x < aT$, the function y in (2.6.7) is a maximum at $x = X$. The value of X can be calculated from (2.6.9) with the aid of an appropriate numerical algorithm. Also note from Figure 2.6.1 that if aT increases, then X also increases.

To summarize the interpretation and implementation of the model (note there is no validation in this case): Using (2.6.6) to return to the original problem, we find that if $aT \leq 2$, then $k = 0$ produces the largest total dividend over the period of T years, which means that all the profit is paid out as a dividend. Hence, it does not pay to re-invest money in the company, because either a or T or both are too small. On the other hand, if $aT > 2$, there exists a unique number $K = \frac{X}{aT}$ such that $Ku(t)$ must be re-invested.

- Read: [7] page 20 also discusses briefly a similar problem. A quick look at [2] will show many mathematical problems in Economics. In [53] pages 75 to 79 an elementary version of the theory of demand is discussed.

- Do: Exercises 16, 17 in §2.12.

- Do: Projects E^* and F in §2.13.

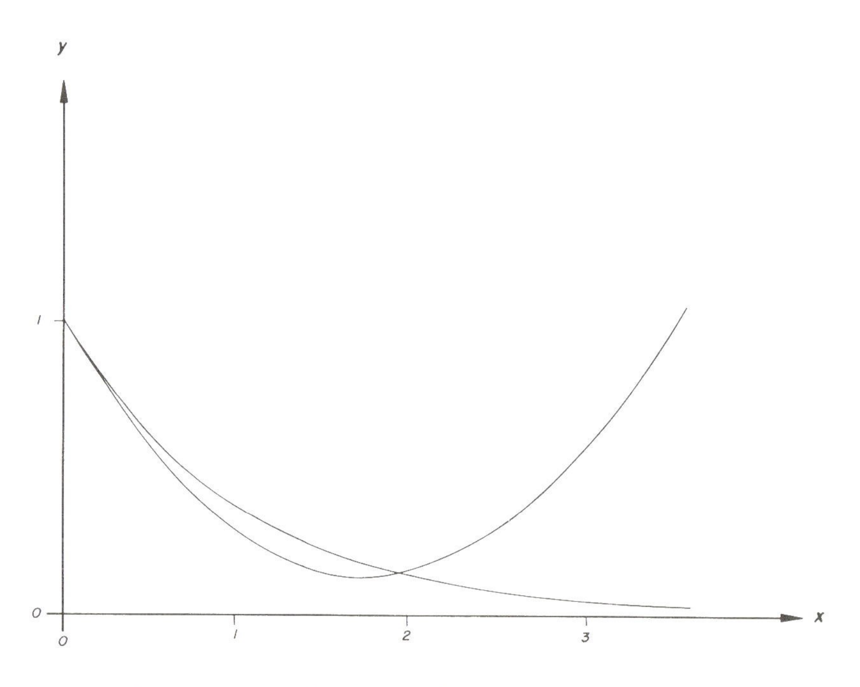

Figure 2.6.1: **Graphs of the functions in (2.6.9)**

2.7 Epidemics

Suppose a community is struck by an epidemic. The medical authorities will try to contain the spread of the disease, if possible. The important issue is how many people will contract the disease if no restrictive measures (quarantine, hospitalization, etc.) or preventative treatment (injections, inoculations, etc.) are introduced. Once this is known, a plan of action can be devised, depending on the seriousness of the disease, and the impact of the intended action can be investigated. In this section we shall construct a simple model of the spread of an epidemic - in fact we have already completed the first stage of identification of the problem. (See Figure 1.1.1).

We shall assume that the underlying mechanism of the spread of a disease is the physical contact of infected and healthy people. To simplify matters we shall assume that a person is either ill or healthy (no recovery from the disease!) and that the community is closed in the sense that nobody enters or leaves the community in the time period under consideration. For further reference, let us formally state:

Assumption (I)

The disease is spread by contact between ill and healthy members of a closed community and there is no quarantine (i. e. people can mix freely in the community).

To construct the mathematical model, let t denote the time (measured in hours, days, or weeks as the case may be) and let $x(t)$ represent the fraction of the community that is ill at time t. By the assumption that the community is closed, it follows then that $1 - x(t)$ is the fraction of healthy people in the community. Hence, if there are n individuals in the community, then $nx(t)$ are ill and $n(1 - x(t))$ are healthy at time t. If n is large, it is reasonable to assume

Assumption (J)

The derivative of the function $x(t)$ is a continuous function of t for $t > 0$.

From Assumption (I) it follows that the number of people infected with the disease in the time interval δt is directly proportional to the number of contacts between healthy and ill people that occurred in the time interval δt, which in turn is proportional to all the possible contacts between healthy and ill people that could have occurred in the time interval δt. In fact, the probability ρ that an ill person will make contact with a healthy person is the quotient

$$\rho = \frac{\text{contacts between ill and healthy people per time unit}}{\text{all possible contacts between ill and healthy people per time unit}}$$

We shall assume that ρ is constant. Note that ρ is a function of time in general - one would expect ρ to be larger during the day than during the night. All the possible contacts are, of course, the product of the number of ill people with the number of healthy people at time t. In symbols, if δx is the change in $x(t)$ during the time interval δt, we have

$$\delta x = kx(1-x)\delta t$$

with k a positive constant (which includes the probability ρ and total population n) called the *infection rate*. If the initial ratio of ill people is denoted by α, with $0 < \alpha < 1$, the following boundary value problem is obtained:

$$\frac{dx}{dt} = kx(1-x), \qquad x(0) = \alpha \qquad (0 < t < \infty) \tag{2.7.1}$$

This is again a logistic model. As in §2.4 the analysis of (2.7.1) produces the unique solution

$$x(t) = \frac{\alpha}{\alpha + (1-\alpha)e^{-kt}} \tag{2.7.2}$$

Note that as $t \longrightarrow \infty$, $x \longrightarrow 1$ which means that everybody will get ill irrespective of the initial population. This is not surprising since no quarantine was imposed (one would expect that the mere fact that a person is ill in bed would automatically limit the number of contacts with healthy people, unless the illness is something like a common cold which does not restrict the movement of the patient), no preventative inoculations or other medical treatment were implemented, and we assumed that people remained ill forever without recovering. Hence the assumptions give rise to an unrealistic result that nobody will escape the disease. However, more realistic assumptions will remedy this deficiency in our model, as the exercises and the next section will show.

- Do: Exercises 18, 19 in §2.12.

2.8 Potato Blight

In this section we consider another case of an epidemic to show the complications that can arise, and especially the analysis to obtain reasonable answers. This real-life model was constructed by J. E. van der Plank in [63] and analyzed by N. Sauer in [54]. Van der Plank did research on the potato blight fungus and noted the following characteristics of the disease:

(1) Immediately after the plant is infected, a latent period occurs in which the plant is not infectious.

(2) After the latency comes an infectious period when the plant infects other plants.

(3) After the infectious period, the plant remains diseased, but cannot infect other plants.

The disease is spread by spores falling on healthy leaves of a plant. During the latent period the plant is forming its own spores which are dispersed during the infectious period. When all the spores had been dispersed, the plant is not infectious any more, and no new spores are formed again by the same plant. Let the latent and infectious periods be of length p and q, respectively. If $x(t)$ again represents the fraction of the population which is diseased, and we use Assumptions (I) and (J), it follows by the same argument that

$$\frac{dx}{dt} = k[1 - x(t)][x(t - p) - x(t - p - q)] \qquad (2.8.1)$$

The expression in the second pair of square brackets describes the fraction of plants which is infectious at time t, when the fraction of diseased plants past the latent period is $x(t - p)$ and the fraction of diseased plants past the infectious period is $x(t - p - q)$. This equation is more difficult to analyse than (2.7.1), since the right hand side includes the unknown function x at different points in time. This is known as a *time-delay equation.* Note that a mere initial value $x(0) = \alpha$ is not sufficient. We must know the history of $x(t)$ on the interval $-p-q \leq t \leq 0$ before we can investigate the behaviour of $x(t)$ for $t > 0$. We shall also have to amend Assumption (J) in this case, as well as the definition of a solution as was stated in §1.2.

Let us consider the case:

$$x(t) = \begin{cases} 0 & (-p - q \leq t < 0) \\ \alpha & (t = 0) \end{cases} \qquad (2.8.2)$$

First restrict t to the interval $[0, p]$. Then the expression in square brackets in (2.8.1) is zero by (2.8.2), and hence $x(t)$ is a constant in this time interval:

$$x(t) = \alpha \qquad (0 \leq t \leq p) \qquad (2.8.3)$$

Consider the case when $q \leq p$. In the interval $(p,\ p + q]$ we have

$$x(t - p) - x(t - p - q) = \alpha - 0 = \alpha$$

by (2.8.2) and (2.8.3) and so (2.8.1) becomes the linear equation

$$\frac{dx}{dt} = k\alpha(1 - x) \qquad (p \leq t \leq p + q)$$

Using $x(p) = \alpha$ to calculate the integration constant, we find the unique solution

$$x(t) = 1 - (1 - \alpha)e^{-k\alpha(t-p)} \qquad (p \leq t \leq p + q) \qquad (2.8.4)$$

Consider next the interval $[p + q,\ 2p)$, then

$$x(t - p) - x(t - p - q) = \alpha - \alpha = 0$$

Again $x(t)$ is a constant in this time interval and by (2.8.4) we have

$$x(t) = 1 - (1 - \alpha)e^{-k\alpha q} \qquad (p + q \leq t \leq 2p) \qquad (2.8.5)$$

In this manner, by moving from one interval to the next one, $x(t)$ can be determined, provided that the integration does not become too tedious or impossible (in which case numerical algorithms must be used).

- Do: Exercises 20, 21 in §2.12.

Note that the solution obtained above in each interval satisfies Assumption (J), but the solution of (2.8.1), (2.8.2) does not satisfy Assumption (J) for all $t > 0$ because the derivative of $x(t)$ is not continuous at $t = p,\ p+q,\ 2p,\ \ldots$ The reason for these discontinuities is the discontinuity in the initial data in (2.8.2). Hence, we should replace Assumption (J) by a weaker assumption:

Assumption (J′)

The derivative of the function $x(t)$ is a piecewise continuous function of t on the open interval $(0, T)$ with T arbitrarily large.

Moreover, since the derivative does not exist at $t = p,\ p+q,\ 2p,\ \ldots$ we cannot use the usual definition of "solution" (see §1.2). In this case we can amend the definition as follows:

Definition 2.5

A solution *of the problem (2.8.1), (2.8.2) on the closed interval* $[0, T]$ *with* T *arbitrarily large, is a function which is continuous on* $[0, T]$, *whose derivative is piecewise continuous on the open interval* $(0, T)$, *which satisfies (2.8.1) at every point in* $(0, T)$ *where the derivative is defined, and agrees with the initial value prescribed in (2.8.2).* □

Finally, the crucial question in the problem of potato blight is what happens in the end: will all the plants in the potato field be infected? The problem is of course formulated over only one season until the crop is harvested. Let us investigate what happens when $t \longrightarrow \infty$, although the problem is only valid for a finite period of time. At least then we shall know the worst that can happen. Let

$$\beta = \lim_{t \to \infty} x(t) \tag{2.8.6}$$

Note that the method above is of no use to determine β. Since $x(t)$ has a positive slope for $t > 0$ by (2.8.1) and $x(t)$ can never exceed 1 (being a fraction), the limit β must exist. At any time t the function

$$f(t) = x(t - p) - x(t - p - q)$$

is known as a function of t due to (2.8.2) and the above method of finding the solution $x(t)$ in successive intervals. Hence successively, the differential equation (2.8.1) can be seen as

$$\frac{dx}{dt} = kf(t)[1 - x(t)]$$

This is a linear equation (see (2.1.4)) which is separable. Implementing the initial value $x(0) = \alpha$ from (2.8.2), it follows as in (2.1.10) that

$$x(t) = 1 - (1 - \alpha)e^{-k \int_0^t f(\tau)\, d\tau}$$

Recall that $x(t)$ is zero on the interval $[-p-q,\ 0)$ and substitute $\sigma = \tau - p$ and $\sigma = \tau - p - q$ in the integrals

$$\begin{aligned} \int_0^t f(\tau)\, d\tau &= \int_0^t x(\tau - p)\, d\tau - \int_0^t x(\tau - p - q)\, d\tau \\ &= \int_0^{t-p} x(\sigma)\, d\sigma - \int_0^{t-p-q} x(\sigma)\, d\sigma \end{aligned}$$

to obtain the general formula

$$x(t) = 1 - (1-\alpha)e^{-k\int_{t-p-q}^{t-p} x(\sigma)\, d\sigma} \tag{2.8.7}$$

Note that if $x(\sigma)$ is substituted successively in (2.8.7) then we obtain (2.8.3), (2.8.4), and (2.8.5). Let $t \longrightarrow \infty$ in the integral and utilize the mean value theorem (see [16] page 141) with $0 < \theta < 1$ to obtain

$$\lim_{t\to\infty} \int_{t-p-q}^{t-p} x(\sigma)\, d\sigma = \lim_{t\longrightarrow\infty} qx(t-p-\theta q) = q\beta$$

Thus it follows from (2.8.7) that

$$\beta = 1 - (1-\alpha)e^{-qk\beta} \tag{2.8.8}$$

Note that β does not depend on the latent period p. For a given q, k, and α the value of β can now be calculated by a numerical procedure like the secant method or the Newton-Raphson method (see §1.4).

- Read: [63] pages 120 to 122, [10] page 55, and [41] pages 356 to 362. The same model applies to the spread of a rumor as in [31] page 14 or [61] pages 232 to 234.
- Do: Exercises 22, 23 in §2.12.
- Do: Project G^* in §2.13.

2.9 Free Fall with Air Resistance

Forces which resist motion play an important part in everyday life. Brakes in cars, friction in many forms, and damped vibrations are commonplace examples. In this section we consider air resistance. It was established by experiment that the magnitude of the force of resistance in air is directly proportional to the speed for low speeds (up to about 25 meters per second), but for higher speeds (up to about 300 meters per second) it is directly proportional to the square of the speed. If the speed increases to the speed of sound, the force of resistance increaes dramatically and cannot be expressed as some power of the speed. This is called the *sound barrier* which must be overcome. In the supersonic range (speeds of more than 340 meters per second) the magnitude of the resistance is again directly proportional to the speed.

The notion of *terminal velocity* plays a central role in a motion which is opposed by some forces. Let us first express such a motion in symbols (the construction stage in Figure 1.1.1).

Suppose a particle describes a rectilinear motion. (The term "particle" just means that the motion of a body is considered as an entire unit without any regard for the dimensions of the body or rotation about its center of mass. The particle is situated at the center of mass of the body.) At any given instant in time t the particle will occupy a certain position on the straight line. To define the position P of the particle, choose a fixed origin O on the straight line and a positive direction along the line. Measure the distance from O to P and record it with a plus or minus sign as y, where the plus or minus sign is assigned according to whether P is reached from O by moving along the line in a positive or negative direction. The number y completely defines the position of the particle and is called the *position coordinate* or the *displacement* of the particle. Once y is known as a function of t, the motion of the particle is known. The *velocity* of the particle is given by

$$v = \frac{dy}{dt}$$

which can be negative or positive. In the case of rectilinear motion the negative or positive sign is used to indicate the direction of the velocity; the absolute value of v is called the *speed* of the particle. The *acceleration* of the particle is given by

$$a = \frac{dv}{dt} = \frac{d^2y}{dt^2}$$

which can also be negative or positive. Sometimes the words *deceleration* or *retardation* are used when a is negative, because the speed then decreases with increasing time.

When several forces act simultaneously on a particle, the vector sum of these forces is called the *resultant force*. The relation between the acceleration of a particle and the forces acting on the particle was formulated in Isaac Newton's famous work "Principia", published in 1687. For our present purposes this relation can be stated as follows:

Newton's second law of mechanics

If the mass of a particle remains unchanged during a motion, then the acceleration of the particle is proportional to the magnitude of the resultant force acting on the particle, and in the direction of this resultant force.

The system of units of force, mass, length, and time is now chosen to give this law the simple form:

$$\text{mass} \times \text{acceleration} = \text{resultant force} \tag{2.9.1}$$

This holds in the metric (or SI) system if force is measured in newton, mass in kilogram, length in meter, and time in seconds; or in the British system if force is measured in poundal, mass in pounds, length in feet, and time in seconds. (See [43] p. 578.)

Let us now use Newton's second law to construct a model of a rectilinear motion due to the constant force F and a force of resistance $P(v)$ (where $v(t)$ denotes the velocity of the particle at time t) acting in opposite directions on a particle of mass m. By (2.9.1) we obtain

$$m\frac{dv}{dt} = F - P(v) \tag{2.9.2}$$

Note that the resistance P opposes the motion and hence the negative sign.

Assume that $P(0) = 0$, that P satisfies a Lipschitz condition, and that $P(v)$ increases indefinitely as v increases. Then the acceleration will decrease as v increases, since F is constant.

Consider the case when the initial value is zero: $v(0) = 0$. Since F is a positive constant, the acceleration will be positive initially. For $t > 0$ the velocity $v(t)$ increases, and consequently $P(v)$ also increases, so that the resultant force decreases until it becomes zero. Hence, v will increase until a speed V is reached where

$$P(V) = F \tag{2.9.3}$$

Thus as v tends to V, the acceleration tends to zero. We call V the *terminal velocity*. Note that if $v(0) = V$, then $v(t) = V$ is a solution of (2.9.2). Hence, by the uniqueness of the solution (see Theorem 2.4.1) as in §2.5, it follows that v cannot exceed V. (By the same argument, this is, of course, also true if the initial speed is any positive number less than V.)

On the other hand, if the initial speed is more than V, then $P(v(0)) > F$ which results in a retardation until the speed again reaches V, by the same argument as above. Hence, the particle will finally move at the terminal speed V, irrespective of the initial speed. This means that we cannot determine the height from which a hailstone or raindrop has fallen by merely measuring their final speeds, because the speed when it reaches the ground only approximates the terminal speed. Think of the immense damage that hailstones or meteorites would inflict if there were no air resistance and hence no terminal speed! (Of course, if there were no atmosphere, there would not be any hailstones either!)

To illustrate the concept of terminal velocity, let us construct a mathematical model for the following problem:

A man drops from an aeroplane at an altitude of 9577 meters with a stopwatch in his one hand and an altimeter in the other. When the altimeter registers 640.5 meters, he stops the watch and opens his parachute. The time taken for the free fall is 116 seconds. Determine the distance that the man had fallen during the free fall at any time instant.

The identification of the mathematical model is clear. For the next stage (see Figure 1.1.1) we assume

Assumption (K)

The air resistance is directly proportional to the square of the speed at any time instant.

We shall also assume that the velocity of the man and the aeroplane was zero when the man left the aeroplane, so that a vertical motion resulted (no horizontal component in the velocity or displacement of the motion of the man).

We can now construct the model. Choose a vertical axis (the y-axis) downwards with the origin at the point of exit from the aeroplane and take the time t as zero at this point. Denote by $v(t)$ the velocity at time t.

Using Newton's second law (2.9.1) and dividing by the mass of the man, we obtain

$$\frac{dv}{dt} = g - kv^2$$

where k is a constant and g is the gravitational acceleration of the earth. Before we analyse the model, let us introduce the terminal velocity V:

$$kV^2 = g \tag{2.9.4}$$

and replace k by V in the differential equation:

$$\frac{dv}{dt} = \frac{g(V^2 - v^2)}{V^2}, \qquad v(0) = 0 \tag{2.9.5}$$

This equation is separable (see (2.1.9)). By partial fractions, or the transformation $v = V \sin\theta$, or using the first integral in the Table of Integrals in Chapter 8 it follows that

$$v(t) = V \tanh \frac{gt}{V} \tag{2.9.6}$$

(The definitions of the hyperbolic functions can be found in §2.14.) The uniqueness of the solution follows by Theorem 2.4.1. Since the velocity is known, we need only to integrate (2.9.6) to find the distance:

$$y(t) = \frac{V^2}{g} \ln\cosh \frac{gt}{V} \tag{2.9.7}$$

Once V is known, the question of the distance of the free fall can be answered. To determine V, use the data given that the man fell 8936.5 meters in 116 seconds. Hence

$$8936.5 = \frac{V^2}{9.81} \ln\cosh \frac{1137.96}{V}$$

which must be solved by an appropriate numerical procedure. Before you start using the secant or the Newton-Raphson method (see §1.4), note that

$$\begin{aligned} \ln\cosh x &= \ln \frac{1}{2} e^x (1 + e^{-2x}) \\ &= x - \ln 2 + \ln(1 + e^{-2x}) \\ &= x - \ln 2 + (e^{-2x} - \frac{1}{2} e^{-4x} + \frac{1}{3} e^{-6x} - \dots \end{aligned}$$

where the Maclaurin expansion of the logarithmic function (see §2.14) is utilized. Take the first two terms as an approximation of $\ln\cosh x$. Then we must solve

$$8936.5 = \frac{V^2}{9.81}(\frac{1137.96}{V} - \ln 2)$$

$$V^2 = 1641.84V + 126485.44 = 0$$

$$V = 81.04 \text{ or } 1560.80$$

The initial velocity of the man was zero, and the velocity increases to the terminal velocity. Hence, the velocity cannot exceed 81.04 meters per second, and the value 1560.80 has no physical meaning in the context of this model. We have obtained this answer by an approximation of $\ln\cosh x$. The error is the series in brackets above. This series is alternating and monotonic decreasing so that its sum is less than the first term (see (2.14.7) in §2.14). Hence

$$\text{Error} < e^{-2x} \approx e^{-28} \approx 10^{-12}$$

Certainly our approximation will not affect the second decimal in the answer of $V = 81.04$ meters per second! If we substitute this value of V in (2.9.7), we can answer the problem posed:

$$y(t) = 669.47 \ln\cosh(0.121t) \tag{2.9.8}$$

Since $y(t)$ is an increasing function of t, this model produces an answer which at least agrees with the real-world situation of a falling body. One also has some confidence in the model, since it is based on one of the laws of mechanics. Thus, at the stage of interpretation we are satisfied that the assumptions are realistic.

Finally, the model can be validated by experimental results. In [14] page 254 an experiment is described where six men jumped from altitudes varying between 3200 and 9600 meters to about 640 meters before opening their parachutes. Each man was equipped with a barograph which recorded his altitude continuously as he fell. The average mass of each man with his equipment was 118.6 kilograms. It was found that four factors affected the results, namely

- the altitude (or equivalently the air density)
- the mass of the man
- the position of the body, and
- the amount of rotation of the body in the free fall.

In Figure 2.9.1 we show approximately the average values of the experimental results (shown as dots) as well as the theoretical curve

$$H(t) = 9577 - y(t)$$

with $y(t)$ as in (2.9.8).

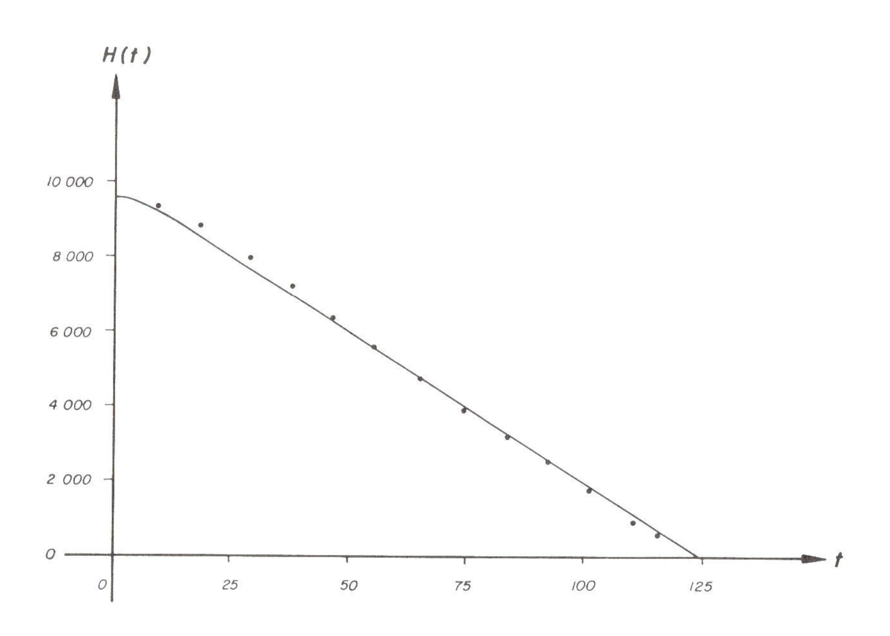

Figure 2.9.1: **Validation of the model for free fall**

The shape of the curve in Figure 2.9.1 suggests that the velocity of the falling body is very quickly close to the terminal velocity. By (2.9.5) we have

$$v(t) = 81.04 \tanh(0.121t)$$

When we calculate v as a fraction of V during the first half minute, we find

t in seconds	$\tanh(0.121t)$
5	0.541
10	0.837
15	0.948
20	0.984
25	0.995
30	0.999

For all practical purposes the body is at the terminal velocity within half a minute. In this half minute the man falls almost 2 kilometers (1966 meters, to be precise).

So when you jump out of an aeroplane and your parachute fails to open, you have good news and bad news: the good news is that your velocity will not exceed the terminal velocity, but the bad news is that the terminal velocity is about 292 km/h or 181 mi/hr!

- Read: [18] page 60, [37] page 24, [50] page 68, [62] page 143 - all describe the situation of free fall.
- Do: Exercises 24, 25, 26 in §2.12.
- Do: Project H^* in §2.13.

2.10 Power

When a machine is evaluated, the rate of doing work is important. Anyone interested in the performance of cars knows about horsepower or kilowatts. To describe this in mathematical terms, the notion of *power* is used. (See, for example, [43] pages 315 and 527.)

Definition 2.6

If a force $\boldsymbol{F}$ causes a particle to move at a velocity $\boldsymbol{v}$, then the power is the scalar product of the two vectors

$$Power = \boldsymbol{F} \cdot \boldsymbol{v} \tag{2.10.1}$$

If $\boldsymbol{v}$ is measured in meters per second and $\boldsymbol{F}$ in newtons, then the power is given in watts. Let us consider a typical problem:

A ship of mass M kilograms is propelled by machines which generate a constant power of H watts. The magnitude of the resistance to the forward motion of the ship is proportional to the speed of the ship at any time. If the ship starts from rest and attains a terminal speed of V meters per second when travelling in a fixed direction, determine the distance travelled until the speed of the ship is $\frac{1}{2}V$, as well as the corresponding time.

Choose the line of travel as the x-axis with the origin at the point of departure and take the time t as zero at this point. Denote by $v(t)$ the velocity at time t. Note that the sign of v determines the direction of the velocity; hence, we shall write the one-dimensional vector $\boldsymbol{v}$ simply as v, and the scalar product reverts to the usual product of real numbers.

If the force of the machines is F newtons at a velocity of v meters per second, then we have by (2.10.1) that

$$F(v) = \frac{H}{v} \text{ newtons}$$

By Newton's second law (2.9.1) the mathematical model is

$$\left.\begin{aligned} M\frac{dv}{dt} &= \frac{H}{v} - kv \qquad (0 < t < \infty) \\ v(0) &= 0 \end{aligned}\right\} \tag{2.10.2}$$

where k is a constant. At the terminal velocity V there is no acceleration, and hence

$$\frac{H}{V} = kV.$$

Solve for k from this equation and substitute in (2.10.2) to obtain an equation in terms of V:

$$\left.\begin{aligned} \frac{dv}{dt} &= \frac{H}{MV^2}\,\frac{V^2 - v^2}{v} \qquad (0 < t < \infty) \\ v(0) &= 0 \end{aligned}\right\} \tag{2.10.3}$$

Since this equation is separable (see (2.1.9)), the solution is easily found by integration (see the third integral in Chapter 8):

$$t = \frac{MV^2}{2H} \ln \frac{V^2}{V^2 - v^2} \tag{2.10.4}$$

This settles the existence of a solution. The uniqueness of the solution cannot be determined by Theorem 2.4.1, because the conditions of this theorem are not satisfied in this case. (If you are not sure why not, read again the definition of a Lipschitz condition in §2.4!) One could, however, rewrite (2.10.3) as

$$\frac{d}{dt}\left(v^2\right) + \frac{2H}{MV^2}v^2 = \frac{2H}{M}$$

which is linear in v^2, and hence, the solution for v^2 is unique and a continuous function of the initial value, by Theorem 2.4.1. From this we obtain two solutions for v when the square root is taken. Note that (2.10.3) is satisfied by both these functions which differ only in sign. In the context of this problem (interpretation stage!), however, v is always non-negative, so that v is then a unique solution to the problem and a continuous function of the initial value.

The second question of the problem can now be answered. The time taken to reach a velocity of $\frac{1}{2}V$ is by (2.10.4)

$$t = \frac{MV^2}{2H} \ln \frac{4}{3} \text{ seconds.}$$

The first question of the problem deals with a relation between speed and distance. There is a standard trick to eliminate t in the derivative in (2.10.3):

$$\frac{dv}{dt} = \frac{dv}{dx}\frac{dx}{dt} = v\frac{dv}{dx}$$

where x denotes the distance covered at time t. Now (2.10.3 can be rewritten as the separable equation

$$\left.\begin{aligned} &\frac{dv}{dx} = \frac{H}{MV^2}\frac{V^2 - v^2}{v^2} \qquad (0 < x < \infty) \\ &v(0) = 0 \end{aligned}\right\} \tag{2.10.5}$$

Using partial fractions or the transformation $v = V \sin\theta$, it follows that

$$x = \frac{MV^2}{H}\left(\frac{V}{2}\ln\frac{V+v}{V-v} - v\right) \tag{2.10.6}$$

The question of uniqueness can be settled by rewriting the differential equation as

$$\left.\begin{aligned} &\frac{dx}{dv} = \frac{MV^2}{H}\frac{v^2}{V^2 - v^2} \qquad (0 < v < V) \\ &x(0) = 0 \end{aligned}\right\}$$

(See [16] page 207 for the reciprocal of $\frac{dv}{dx}$.) Substitute $v = \frac{1}{2}V$ in (2.10.6) to obtain

$$x = \frac{MV^3}{2H}(\ln 3 - 1)$$

- Do: Exercises 27, 28, 29, 37 in §2.12.

2.11 Rockets

When a car is driven over a long distance, the mass of the car changes as the gasoline in the tank is consumed. The change in mass is so small in comparison to the total mass of the car, that one would assume in the corresponding mathematical model that the mass of the car remains constant, even though it is not strictly true. On the other hand, if the motion of a rocket after blast-off is considered, the mass diminishes substantially as the rocket rises. This calls for a new model because Newton's second law of mechanics (see (2.9.1)), which we had used so far to model motion, is not valid in this case.

We defined the notions of displacement, velocity, and acceleration of a particle in §2.9. We now also need the notion of *momentum* of a particle.

Definition 2.7

The linear momentum (or momentum) of a particle is the product of the mass and the velocity of the particle.

Momentum is used, for example, to describe an impact. When a ball hits you, it is not only the mass of the ball which is important, but also the velocity at which it was thrown. (Would you prefer a baseball at 10 meters per second or a golfball at 25 meters per second?) Momentum was also used in the original version of Newton's second law of mechanics. In this form the law states that:

Rate of change of momentum = resultant force (2.11.1)

("Rate of change" just means the derivative with respect to time.) Note that if the mass is constant then (2.11.1) is identical to (2.9.1) . If the mass is not constant, then *this law is only valid when the momentum of the mass added or lost at any time is zero.* This is, for example, the case when a snowball rolls down a slope covered in snow or when a raindrop accumulates moisture as it passes through a cloud. In the case of a rocket, the mass is lost by exhaust gases streaming out of the engines, so that this law is not applicable to the rocket in isolation. One could circumvent this by considering the rocket and exhaust gases as a unit, as was done in [10]. Since the speeds of the gas particles are all different at each time instant, it is burdensome to calculate the total momentum of the system.

We shall use a simpler approach to model rocket motion. For this we need the notion of *impulse*.

Definition 2.8

The impulse of a force $\boldsymbol{F}$ *during a given time interval* $a \leq t \leq b$ *is defined as the integral*

$$\int_a^b \boldsymbol{F}\, dt$$

For example, the shock up your arm when you hit a fast ball in baseball is due to the impulse transmitted to your hand. The amount of damage when a car hits a wall depends on the impulse received by the car. An impulse need not be short, for example, when a lever is pulled.

You may have the feeling that there must be some relationship between momentum and impulse. For one thing, their dimensions are the same. If one integrates Newton's second law of mechanics (2.11.1), the very important *impulse-momentum law* is obtained. This law states that

Impulse-momentum law

The impulse of the resultant force acting on a particle over a given time interval is equal to the change in momentum of the particle in the same time interval.

(See [24] page 50, [37] page 163, or [43] page 111.) We can summarize this law briefly as

$$\text{Impulse} = \text{change in momentum} \tag{2.11.2}$$

We shall now use this law to derive an equation which will enable us to model rocket motion.

Consider a variable mass $m(t)$ moving at a velocity $\boldsymbol{v}(t)$ at time t. In the time interval δt the mass and velocity change to

$$\begin{aligned} m(t+\delta t) &= m(t)+\delta m \\ \boldsymbol{v}(t+\delta t) &= \boldsymbol{v}(t)+\boldsymbol{\delta v} \end{aligned}$$

The amount of mass δm being added may have a velocity relative to the mass $m(t)$. Denote this relative velocity by $\boldsymbol{u}(t)$; the velocity of $m(t)$ is thus $\boldsymbol{u}(t)+\boldsymbol{v}(t)$ at any time instant t.

$$\begin{aligned} &\text{Change in momentum in the time interval } \delta t \\ &\quad = m(t+\delta t)\boldsymbol{v}(t+\delta t) - [m\boldsymbol{v}(t) + \delta m(\boldsymbol{u}(t)+\boldsymbol{v}(t))] \\ &\quad = m\boldsymbol{\delta v} + \delta m\boldsymbol{\delta v} - \delta m\boldsymbol{u}(t) \end{aligned}$$

Denote by $\boldsymbol{F}(t)$ the total external force (the resultant force) on $m(t)$, and assume that $\boldsymbol{F}(t)$ is a continuous function of t. Then

$$\text{Impulse} = \int_t^{t+\delta t} \boldsymbol{F}(\tau)\, d\tau = \boldsymbol{F}(\xi)\delta t$$

with $t < \xi < t+\delta t$ by the mean-value theorem (see [16] page 141). Now we implement the impulse-momentum law (2.11.2) to obtain

$$\boldsymbol{F}(\xi)\delta t = m\boldsymbol{\delta v} + \delta m\boldsymbol{\delta v} - \delta m\boldsymbol{u}(t)$$

Divide by δt and let $\delta t \to \infty$ to obtain

$$\boldsymbol{F}(t) = m\frac{d\boldsymbol{v}}{dt} - \boldsymbol{u}\frac{dm}{dt} \qquad (2.11.3)$$

Note that if $\boldsymbol{u} = -\boldsymbol{v}$ with the implication that the velocity of the particle which joins $m(t)$ is zero, then (2.11.3) reduces to (2.11.1). Hence, we have deduced from the impulse-momentum law a more general law than Newton's second law of mechanics.

You are now ready to tackle a problem about a rocket:

A rocket of mass M kilograms is fired vertically upwards from the launching pad. The engines blow exhaust gases out at a constant rate of α meters per second, while the fuel is consumed at a constant rate of β

kilograms per second. At the start the fuel in the rocket is nM kilograms where $0 < n < 1$. Determine the altitude and velocity of the rocket at any time t during the flight.

Before the model can be constructed, we must first make three important assumptions:

Assumption (L)

The motion of the rocket is resisted by a force whose magnitude is directly proportional to the speed at any time during the flight.

Assumption (N)

The maximum altitude of the rocket is small enough so that the acceleration of gravity remains the constant g during the flight.

Assumption (O)

The gases leaving the exhaust nozzle of the rocket are at atmospheric pressure.

Let us start with the construction stage of the modelling process. Choose a y-axis vertically upwards with the origin at the launching pad. Denote, respectively, by $y(t)$, $v(t)$, and $m(t)$ the altitude (in meters), the velocity (in meters per second), and the mass (in kilograms) at time t where $t = 0$ at the moment of blast-off. As before, the sign of v will suffice to describe the direction of v so that we shall not write $\boldsymbol{v}$. By the assumptions the total external force on the rocket is

$$\boldsymbol{F}(t) = -m(t)g - k\beta v(t)$$

where k is a positive constant and β is merely included to simplify subsequent formulas. The direction of the resultant force $\boldsymbol{F}$ is, of course, always downwards as the rocket rises. By (2.11.3) it now follows that

$$m\frac{dv}{dt} - (-\alpha)\frac{dm}{dt} = -mg - \beta kv \tag{2.11.4}$$

since the relative velocity $\boldsymbol{u}$ of the exhaust gases is, in this case, in the opposite direction as $\boldsymbol{v}$. We also know that

$$\frac{dm}{dt} = -\beta \tag{2.11.5}$$

where the minus sign shows that $m(t)$ decreases as t increases. Let T denote the time when all the fuel has been consumed, then by (2.11.5) we have

$$m(t) = \begin{cases} M - \beta t & (0 \leq t \leq T) \\ (1-n)M & (t \geq T) \end{cases} \tag{2.11.6}$$

Let $t = T$ in this equation, then it follows by the continuity of $m(t)$ that

$$T = \frac{nM}{\beta} \text{ seconds} \tag{2.11.7}$$

For the analysis of the model, consider first the motion in the time interval $0 \leq t \leq T$. By (2.11.4) and (2.11.6) we obtain the initial value problem

$$\left.\begin{aligned} &\frac{dv}{dt} + \frac{\beta k}{M - \beta t} v = \frac{\alpha\beta}{M - \beta t} - g \quad (0 < t < T) \\ &v(0) = 0 \end{aligned}\right\} \tag{2.11.8}$$

This differential equation is linear in v with the integrating factor (see (2.1.13))

$$P = e^{\beta k \int \frac{dt}{M - \beta t}} = (M - \beta t)^{-k}.$$

After all the integrals are computed (see (2.1.14)), we finally find that

$$v(t) = \frac{\alpha}{k} + \frac{g}{\beta(1-k)}(M - \beta t) - \left(\frac{gM}{\beta(1-k)} + \frac{\alpha}{k}\right)\left(1 - \frac{\beta t}{M}\right)^k \tag{2.11.9}$$

provided that $k \neq 1$. (In actual fact with air resistance $k \ll 1$.) If (2.11.9) is integrated again, we obtain

$$\begin{aligned} y(t) = \frac{\alpha t}{k} + \frac{gt}{2\beta(1-k)}(2M - \beta t) \\ - \frac{M}{\beta(1+k)}\left(\frac{gM}{\beta(1-k)} + \frac{\alpha}{k}\right)\left[1 - \left(1 - \frac{\beta t}{M}\right)^{k+1}\right] \end{aligned} \tag{2.11.10}$$

where (2.11.9) and (2.11.10) are only valid for $0 \leq t \leq T$.

When $t > T$ the engines have stopped and the rocket becomes a free falling body. This case was studied in §2.9 with Assumption (K) instead of the easier Assumption (L). Hence, you are now able to calculate the maximum altitude and the total flight time of the rocket. The case when the rocket is fired at an angle less than 90° (in other words not vertically upwards) is more difficult. We shall meet this type of problem in Chapter 5.

It is interesting to note that if the rocket is designed to rise immediately after ignition, we must have

$$\left.\frac{dv}{dt}\right|_{t=0} \geq 0.$$

By (2.11.8) this means that

$$\frac{\alpha\beta}{M} - g \geq 0$$

or equivalently

$$\alpha\beta \geq Mg \tag{2.11.11}$$

If this condition is not met, then the rocket will stay on the launching pad until enough fuel has been consumed (in other words the mass has decreased) to make

$$M - \beta t \leq \frac{\alpha\beta}{g}$$

unless, of course, after T seconds the mass is still too heavy, in which case there is no lift-off at all, and many red faces! (Do not worry - it cannot happen to you, having taken this course!)

- Read: [10] page 82 where (2.11.1) is used to obtain the equation for the rocket by considering the rocket and the exhaust gases as a unit. Rockets are also discussed in [37] page 33, [50] page 76, [58] page 113, [62] page 191, [43] page 557.
- Do: Exercises 30, 31 in §2.12.
- Do: Project I^* in §2.13.

2.12 Exercises

(1) The bacteria in a culture numbered 10,000 initially. Two and a half hours later the bacteria had increased by 10%. Assume that the rate of change is directly proportional to the number of bacteria.

(a) How many bacteria will there be after 10 hours?

(b) How long will it take until the bacteria numbers 20,000?

(2) The population of a city is initially 40,000. The net growth rate due to births and deaths is such that the population will double in 50 years. The city also gains every year 400 people who come to settle in the city. Construct a simple model to predict the population of the city after 10 years.

(3) The rate of change in a colony of ants is directly proportional to the population. The population doubles in 2 weeks. If all the eggs are removed so that no births take place, then only half the population is left after 1 week. Calculate the birth rate as a percentage per week.

(4) A colony of ants is part of a biological experiment in a laboratory. The estimated size of the population in the colony is given in the following observations:

t (in days)	N (in thousands)
0	4.0
1	5.0
2	6.5
3	8.5

Use the assumptions (A) and (B) in §2.2 to construct a mathematical model for the population growth in the colony.

(a) Utilize then the method of least squares to calculate the two constants in the model.

(b) Predict the population on the seventh day.

(5) The well-known fruitfly *Drosophila gansobscura* is intensively studied in the laboratory to determine its population growth and especially the effect of different kinds of poison on the population. For the normal population growth the following estimated population is observed:

time (in days)	population (in hundreds)
0	5.0
1	6.0
2	7.5
3	9.0

Use the assumptions (A) and (B) in §2.2 to construct a mathematical model for the population growth of the fruitfly.

(a) Apply then the method of least squares to calculate the two constants in the model.

(b) Predict the population on the seventh day.

(6) State whether each of the following functions satisfies a Lipschitz condition or not on the given interval

(a) $f(x) = 3x^2 + 5 \quad (0 \leq x \leq 6)$
(b) $f(x) = x^{\frac{2}{3}} \quad (-1 \leq x \leq 1)$
(c) $f(x) = \frac{1}{x} \quad (1 \leq x \leq 3)$
(d) $f(x) = |x| \quad (-1 \leq x \leq 1)$

(7) Show that the function $f(x) = \sqrt{1 - x^2}$ is continuous on the closed interval [-1;1], but does not satisfy a Lipschitz condition on [-1;1].

(8) Show that the function

$$f(x) = \begin{cases} 1 + x & (-1 \leq x \leq 0) \\ 1 - x & (0 \leq x \leq 1) \end{cases}$$

satisfies a Lipschitz condition with $k = 1$.

(9) Two hundred flies are imprisoned in a large bottle with a fixed amount of food at the start of a laboratory experiment. The number of flies alive in the bottle is counted every morning at 9:00, after which the food is replenished for the day. It is known that the birthrate of these flies is 150% per day. After 2 days the count is 360 flies in the bottle.

(a) Use the logistic model to determine the number of flies alive in the bottle 3 days after the experiment started.

(b) Find the limit of this population of flies.

(This exercise will be continued in Exercise 15.)

(10) The model used by Smith was discussed in §2.4.

(a) Solve (2.4.9) with $N(0) = \alpha$ and $\alpha < b$.

(b) What happens when $t \to \infty$?

(c) Draw a qualitative graph of $N(t)$ for the cases $\alpha < b$ and $\alpha > b$. What happens when $\alpha = b$?

(d) Use Theorem 2.4.1 to show that the solution is unique.

(11) In a research project on the population of fish in the North Sea a model of the mass of a fish had to be constructed. If $m(t)$ represents the mass of a fish, the following differential equation was found in [12]:

$$\frac{dm}{dt} = \alpha m^{\frac{2}{3}} - \beta m$$

with α and β positive constants. If the initial mass of the fish is approximately zero, solve the initial value problem and draw a qualitative graph of $m(t)$.

(12) Suppose that $E > F$ in the model of §2.5 and that the initial population is $\alpha = \frac{b}{s}$. Show that (2.5.4) can be written as

$$T = \frac{2}{\sqrt{s(E-F)}} \arctan\left(\frac{b}{2\sqrt{s(E-F)}}\right)$$

Calculate T when $E = 10,000$ birds per year and b and s have the values in (2.5.5).

(13) Hake is intensively fished in the Atlantic Ocean in the vicinity of the Cape of Good Hope. Construct a model for the biomass of the hake population with assumptions like Assumption (C) and Assumption (D) of §2.4, and with the additional assumption that the annual catch is directly proportional to the biomass in that year. Given the following data

Year	Biomass in metric tons
1955	319,000
1964	473,000
1973	685,000

(a) calculate the constants in the model, and

(b) use the model to predict the biomass in 1995.

(c) What will happen to the biomass when $t \to \infty$?

(14) In a wildlife reserve a non-indigenous bird endangers the other birds. To control this pest a scientific study is made of the population growth of the unwelcome visitors, and it is found that

$$\frac{dN}{dt} = \frac{(9 - 0.00005N)N}{100 + 0.02N}$$

where $N(t)$ represents the estimated population and t is measured in years. The manager of the reserve is considering allowing controlled hunting of E birds annually to limit this pest.

(a) Determine the critical value F of harvesting. Show that if $E > F$ then the population will be zero in a finite time.

(b) If $E = 250$ birds per year, draw a qualitative graph of N as a function of t to show the different possibilities for different initial populations.

(c) If $E = 475$ birds per year and the initial population is 30,000 birds, calculate the extinction time T for these birds.

(15) Exercise 9 continued:

(a) If 80 flies are removed daily from the bottle, determine the limit of the population in this case.

(b) Determine the critical number of flies which must be removed each day to ensure that the population will die out, but if one fly less is removed each day, the population will not die out.

(16) The capital investment of a company is α initially. An amount of b times the time interval must first be subtracted from the profit over the same time interval before the remaining profit can be divided between re-investment and a dividend for the shareholders, as in §2.6. Here b represents a fixed annual fee for the members of the board. Construct a model similar to the model in §2.6. State explicitly any additional assumptions that you may deem necessary. (Note that $b = 0$ in your model must reproduce the model in §2.6. Use this fact as a check on your work.)

(17) The capital of a company is \$500,000. The expected profit of the company is estimated at 14% of the capital per annum over a period of 20 years. The board decides to re-invest a fraction k of the profit at time t as capital in the company. After 40% of the profit is also deducted as tax, the remaining profit is paid out as a dividend to the shareholders. Assume that the capital is a continuous function of time to construct a model for the total dividend over the period $[0, t]$. Solve the model and determine the value of k which maximizes the total dividend over a period of 15 years. Calculate this optimal dividend over 15 years.

(18) Construct a model for an epidemic, as in §2.7, except that in this case the medical officer starts a program for inoculation at time $t = 0$. A person who has been inoculated is immune to the disease. At time t the number of healthy people inoculated is directly proportional to the number of people who had been infected, with the constant of proportionality a (in other words the rate of inoculation depends on the rate at which the disease spreads). Note that a must satisfy $\alpha + a\alpha \leq 1$. Use your model to answer the following two questions:

(a) How many people will become ill in the end?

(b) Determine the value of the constant a which will ensure that less than half the population will become ill.

(19) The manufacturers of a new brand of soda *Downtwo* decide to advertise their product in the remote town of Thirstville by sending

free miniature bottles to a quarter of the population. The manufacturers assume that these people will tell the rest of the population about *Downtwo*. Construct a model which describes the spread of the news about *Downtwo*, and solve it. If the manufacturers expect half the population to know about *Downtwo* after one week, calculate when $\frac{4}{5}$ of the population will be informed. (Hint: use Assumptions (I) and (J) of §2.7.)

(20) In the problem of potato blight in §2.8, continue the solution in (2.8.3), (2.8.4), and (2.8.5) to the next interval $2p \leq t \leq 2p + q$ (to convince you that matters can get quite messy quickly!).

(21) The town Slanderville (population 5000) is well-known for the speed at which a rumour can spread amongst the inhabitants. In a sociological research project it is found that when a person hears a rumour, he (or she!) will tell it to everybody for a period of T days until the person loses interest. On a certain Saturday afternoon a bus returned to Slanderville with fifty passengers aboard. On the way home they saw a nasty car accident happen before their very eyes, and they could not wait to tell everybody about it. Construct a model to describe the spread of this news in Slanderville. Take care to formulate all your assumptions. Show how you will solve this model.

(22) In Exercise 21 if the constant of proportionality $k = 0.5$ and $T = 2$ days, calculate the number of people in Slanderville who would have heard of the accident as the time $t \to \infty$.

(23) Draw a graph of a straight line and of an exponential function on the same β−axis to determine whether β, as given in (2.8.8), will increase or decrease as kq increases. Show also on the graphs that if $\alpha \neq 1$ we must have that $\beta < 1$; in fact $\beta < 1 - (1 - \alpha)e^{-qk}$.

(24) The mass of a man equipped with a parachute is 100 kilogram. He drops from an aeroplane flying horizontally and waits for 10 seconds before opening his parachute. If it is known that the terminal speed of a free falling body is 290 km/h and if it is assumed that the air resistance is directly proportional to the square of the speed at any time, calculate the distance after 10 seconds and the speed at that moment. (Ignore the horizontal movement of the man.)

(25) The terminal speed of a bomb is found to be 400 km/h by experiment. The designers wish the bomb to explode at an altitude of 600 meters after being dropped from 10,000 meters in horizontal flight. At how many seconds should the time delay of the firing

mechanism be set if the mechanism is activated when the bomb leaves the plane? To simplify the model, ignore the horizontal movement and assume that the air resistance is proportional to the square of the speed.

(26) A gun fires a projectile vertically upward at W meters per second. Assume that the air resistance is proportional to the square of the speed. Construct a model to determine the velocity and the distance covered by the projectile at any time. If k is a constant, g the acceleration of gravity, and V the terminal speed, show that

(a) the projectile rises to an altitude of

$$\frac{1}{2k}\ln\left(1+\frac{kW^2}{g}\right) \text{ meters}$$

(b) the time to reach the maximum altitude is

$$\frac{\arctan\left(\sqrt{\frac{k}{g}}W\right)}{\sqrt{gk}} \text{ seconds}$$

(c) the speed at which the projectile strikes the ground is

$$\frac{VW}{\sqrt{V^2+W^2}} \text{ meters per second}$$

(d) the total time in the air is

$$\frac{V}{g}\arctan\left(\frac{W}{V}\right)+\frac{V}{g}\ln\left[\frac{W}{V}+\sqrt{1+\frac{W^2}{V^2}}\right] \text{ seconds}$$

(We assume that the height of the gun is neglible compared to the maximum altitude of the projectile.)

(27) A machine operates at a constant power p to move a mass m horizontally against a constant resistance c. If the initial speed is zero, determine the time required to reach two thirds of the maximum speed.

(28) An engine pulls a train of mass 300 metric tons on a horizontal track at a constant power of 300 kilowatts. The resistance to the movement consists of the sum of two components: one component remains constant and the other is proportional to the speed of the train. In an experiment it is established that at a speed of 40 km/h the resistance is 11,772 newtons, and at a speed of 24 km/h the resistance is 8829 newtons. Calculate the maximum speed of the train and the time to reach a speed of half the maximum speed if the train starts from rest.

(29) An engine pulls a train of mass 300 metric tons on a horizontal track at a constant power of 480 kilowatts. At a speed of 36 km/h the resistance is measured as 30 newtons per ton. Assume that the magnitude of the resistance is directly proportional to the square of the speed at any time. Calculate the maximum speed of the train.

(30) A rocket is launched in a vertical direction. The total mass is 25,000 kilograms of which 20,000 kilograms is fuel. The engines emit exhaust gases at a constant rate of 400 meters per second and consume fuel at a constant rate of 100 kilograms per second. The resistance is $-2\boldsymbol{v}$ newtons with $\boldsymbol{v}(t)$ the velocity of the rocket (measured in meters per seond) at time t. Assume that the acceleration of gravity is 9.8 m/s^2 during the flight.

(a) Show that the following boundary value problem is valid until all the fuel is consumed:

$$\begin{cases} (250-t)\dfrac{dv}{dt} = 400 - 9.8(250-t) - 0.02v \\ v(0) = 0 \end{cases}$$

(b) Solve the boundary value problem in (a).

(c) Calculate v at the moment when all the fuel had been consumed. What do you infer from your answer as regards the motion of the rocket?

(31) A rocket is to be launched in a vertical direction for a test flight. The total mass is 28,000 kilograms of which 20,000 kilograms is fuel. The engines emit exhaust gases at a constant rate of 500 meters per second and consume fuel at a constant rate of 800 kilograms per second. The designers of the rocket want an estimate of the maximum altitude of the rocket. They decide to ignore air resistance and to assume that the acceleration of gravity will be 9.81 $\mathrm{m/s}^2$ during the flight.

(a) Calculate the maximum altitude of the rocket as given by this simplified model.

(b) Investigate whether the assumption that the acceleration of gravity is constant during the flight is valid in the light of your answer. (Approximate the earth as a sphere with radius 6400 kilometers.)

(32) Solve the boundary value problem (1.3.2), (1.3.3) in §1.3.

(33) (a) Solve the boundary value problem

$$\begin{cases} \dfrac{dy}{dx} = \dfrac{2x(x^2+1)}{y} \\ y(0) = 0 \end{cases}$$

to show that there are two solutions.

(b) Use a computer to draw the graphs of these solutions and compare your results with the direction field of Figure 1.3.3.

(c) Investigate what happens when the boundary value is changed to $y(1) = 0$.

(34) (a) Solve the boundary value problem

$$\frac{dy}{dx} = -\frac{\sin x}{y}$$

with each of the following initial values
(i) $y(0) = 2$ (ii) $y(0) = 1$ (iii) $y(0) = 3$.

(b) Use a computer to draw the solutions and compare your result with the direction field found in Exercise 6(c) of §1.5.

(35) Solve the boundary value problems:

(a) $$\begin{cases} \dfrac{dy}{dx} + 2xy = x \\ y(0) = 1 \end{cases}$$

(b) $$\begin{cases} (x^2+1)\dfrac{dy}{dx} - xy = x(1+x^2) \\ y(0) = 2 \end{cases}$$

(36) Solve the boundary value problems:

(a) $$\begin{cases} (x^4 - 2xy^3)\dfrac{dy}{dx} + y^4 - 2x^3y = 0 \\ y(1) = 1 \end{cases}$$

(b) $$\begin{cases} \dfrac{dy}{dx} = \dfrac{3x-y}{x+y} \\ y(1) = 2 \end{cases}$$

(37) A particle of mass M moves along the x-axis and is attracted towards the origin by a force proportional to its displacement from the origin (as, for example, in the case of a spring oscillating up and down - see later in §6.1). It is also subject to a resistance which is directly proportional to the velocity of the particle. Show that the equation of motion is

$$v\frac{dv}{dx} = -cv - kx$$

where v is the velocity of the particle and c and k are positive constants. If $c^2 < 4k$ determine v as an implicit function of x. (Hint: use the Table of Integrals in Chapter 8 and (2.1.17).)

2.13 Projects

The projects in this section will be more time-consuming than the exercises in §2.12. On the other hand they are shorter than the excellent modules of UMAP. Each project is designed to give the reader personal experience in one or more of the modelling stages of Figure 1.1.1. The ability to write a computer program for a microcomputer (or personal computer) is a prerequisite for the projects marked by an asterisk.

Project A^*: Curve Fitting for the Malthus model

(1) Given a table of data $\{(t_i, N_i) : \ i = 0, 1, \ldots, n\}$, write a program which will calculate the best fit by the method of least squares for a Malthus model and then plot the data and the graph on the same axes. Verify your program with the example in §2.3.

(2) The estimated population of all the people on earth is given in the following table:

Year	World population (in millions)
1650	545
1750	728
1800	906
1850	1171
1900	1608
1950	2509
1960	3010
1970	3611
1973	3860

Try to fit this data to a Malthus model by using your program for the method of least squares. Discuss the validity of the model.

Project B^*: Getting a Grip on Parameters

(1) Given three data points $\{(t_i, N_i) : \quad i = 1, 2, 3\}$ which must be used to calculate the three parameters α, s, and b in (2.4.4), the solution of the logistic model. Substitute the data in (2.4.4) to obtain three equations in the unknowns α, s, and b. Introduce new variables to simplify the analysis: $y = \frac{s\alpha}{b}$, and $x = e^{-b}$.

By elimination show that x satisfies

$$(N_1 - N_2)N_3 x^{t_3 - t_1} + (N_3 - N_1)N_2 x^{t_2 - t_1} = (N_3 - N_2)N_1 \quad (2.13.1)$$

Show also that a solution x of (2.13.1) such that $0 < x < 1$ is only possible if

$$(t_2 - t_1)(N_3 - N_1)N_2 < (t_3 - t_1)(N_2 - N_1)N_3 \qquad (2.13.2)$$

where we assume that $t_1 < t_2 < t_3$ and $N_1 < N_2 < N_3$.

(2) Write a program which will either calculate the parameters α, b and s of (2.4.4) as in (1) above (using the first three data points), or state that the model does not fit the data. The program must also draw the graph of $N(t)$ (if α, b, and s can be found) and plot the given number of n data points (t_i, N_i). Moreover, the program must be able to calculate the sum of the square of the errors at each data point

$$E = \sum_{i=1}^{n} (N(t_i) - N_i)^2$$

(3) Verify your program with the model of Pearl and Read in §2.4.

(4) Try to obtain a better fit (in other words to make E less) by choosing any three data points (instead of the first three) from Figure 2.4.1 in your program.

(5) Try to fit the logistic model to the data given in the table in Project A^*.

Project C: Population Explosion

(1) In the notation of §2.4, assume that the rate of change of the population N is of the form $(bN - s)N$ instead of Assumption (C). This is the case of a population explosion where the death rate remains constant, but the birth rate increases with time. This type of situation occurs, for example, in the population of neutrons in a nuclear reactor. Construct a model to predict the population at any time $t \geq 0$. Draw a qualitative graph of the resulting function $N(t)$ to interpret your solution.

(2) The increase in the population on the planet Earth is matter of grave concern. Pictures of starving people appear regularly in newspapers and on television programs. Some experts predict a catastrophe unless the birth rate is forced to fall from the present high levels. The question arises whether a "population explosion" could result. Investigate this possibility with the model constructed in (1) above. Use the data points at 1650, 1750, and 1850 in the table of Project A^* to calculate the three parameters in the model. Tabulate the predicted world population and the data given in Project A^* to validate the model. Can this model be implemented?

Project D^*: Extinction Times for Endangered Species

In the notation of the model in §2.5 it was found in (2.5.4) that the extinction time T is

$$T = \int_{\alpha}^{0} \frac{dN}{(b - sN)N - E}.$$

Write a program with the following features:

(1) Calculate the critical harvesting rate F.

(2) Check whether $E > F$. If so, continue to calculate T. If not, check whether $\alpha < \beta_2$ (see Figure 2.5.2). If so, continue to calculate T.

If not, inform the user that the population will not die out, and supply the limit β_1.

(3) Use integration methods to obtain expressions for T in the three cases

(a) $E > F$ (see also Exercise 12 in §2.12)

(b) $E < F$ and $\alpha < \beta_2$ and

(c) $E = F$ and $\alpha < \beta$.

(4) Verify your program with the data in §2.5 on the sandhill crane.

(5) Draw a graph of the extinction time T for the sandhill crane on the y-axis and the harvesting rate E on the x-axis, using the data in (2.5.5) and the initial value 194,600.

Project E^*: Planning for a Maximal Dividend

In §2.6 we discussed the long-term planning of the management of an idealized company to ensure an annual dividend which will aggregate to a maximal return on the capital investment over a number of years. In practical terms the model requires the solution of a nonlinear equation (2.6.9) for which a numerical algorithm must be used. The chairman of the board of this company, who happens to be your father, asks you to write a program - a very user-friendly program for a chairman who knows the barest rudiments of a microcomputer. The purpose of the program is to enter the parameters α, a, T and then to calculate k, $u(t)$, and $w(t)$. Since you were on the point of asking him for a loan (interest-free, repayable in the year 2050) this seems to be a golden opportunity to create a benevolent atmosphere.

(1) Write a program to calculate k, using the Newton-Raphson method (see §1.4) to solve (2.6.9) for all reasonable values of a and T. Print the value of k and draw graphs of $u(t)$ and $w(t)$ on the closed interval $[0, T]$.

(2) Implement your program with the following data: The expected profit of the company is estimated at 15% of the capital investment annually for a period of 30 years. The initial capital was 5 million dollars. Consider three cases for T:

(a) 10 years

(b) 15 years

(c) 25 years.

Project F: Marketing

Suppose certain goods are marketed at an initial selling price of α. The changing demand and supply of the goods induce adjustments in the price as time goes on. Let us try to construct a mathematical model which will describe the change in the selling price $p(t)$ with time t, and which could perhaps predict an equilibrium price β (when p remains constant). As before, we must make the basic assumption:

> **Assumption (G)**
>
> *The derivative of the function $p(t)$ is continuous for $0 < t < \infty$.*

It is also reasonable to assume that if the demand exceeds the supply, then the selling price p will rise, and conversely.

> **Assumption (H)**
>
> *The rate of change of the selling price p is directly proportional to the difference between demand and supply.*

To construct a model we need some more assumptions about the relation between demand, supply, and price. We look at more than one possibility.

(1) Let us start with the assumption that the demand and supply depend only on the price p. (This is, of course, not generally true, but will suffice for a first crude model.)

Secondly, we assume that the *demand* $u(t)$ at time t is a monotonic decreasing function of p, with u very large if $p = 0$ and the demand diminishing as p grows.

Thirdly, we assume that the *supply* $v(t)$ at time t is a monotonic increasing function of p, with $v = 0$ until p is large enough to make the production of the goods profitable.

Let us assume simple linear functions to describe u and v, namely

$$\left.\begin{aligned} u(t) &= a - bp(t) \\ v(t) &= c + dp(t) \end{aligned}\right\} \qquad (2.13.3)$$

with a, b, and d positive constants and a larger than the constant c. Note that (2.13.3) does not hold near $p = 0$.

With these assumptions, construct a model which expresses the derivative of p in terms of b, d, and β. Solve the boundary value problem, and check the uniqueness and the continuous dependence on α of the solution. Note that p tends to β irrespective whether α is larger or smaller than β. We then say that the market is *stable*.

(2) Suppose that $u(t)$ is not a linear function as in (2.13.3), but rather an oscillating function of the form

$$u(t) = a - bp(t) + f \sin \omega t$$

with b, f, and ω positive constants. Use this assumption and $v(t)$ as in (2.13.3) to construct a new model. Solve the boundary value problem and investigate the uniqueness of the solution. What happens to $p(t)$ when $t \to \infty$?

(3) We assumed above that the supply and demand depend only on the selling price p. It is more realistic that supply and demand will also be affected by a falling or rising selling price. For example, people will be more inclined to buy an automobile if they knew that the price is rising, and less inclined if they know that the price is falling, because they might get it cheaper later. This is called *price speculation*. To incorporate this into the model let us replace (2.13.3) by

$$\begin{aligned} u(t) &= a - bp(t) + m\frac{dp}{dt} \\ v(t) &= c + dp(t) + n\frac{dp}{dt} \end{aligned}$$

where a, b, c, d, m, and n are constants with b, d, m, and n positive. Note that a rising selling price implies a positive slope of $p(t)$. Let us also assume that the demand and supply are in equilibrium; in symbols $u(t) = v(t)$ for all $t \geq 0$.

(a) Formulate a boundary value problem in this case and determine the solution. Is the solution unique? (What happens when $m = n$?)

(b) Determine the equilibrium price in this case.

(c) Investigate the stability of the market in this case.

Project G^*: Flu in the Bush

In a massive military exercise with 6000 men in a remote corner of the Kalahari in Africa, 6 men suddenly report sick one morning. On examination the medical staff finds that the men had contracted flu. This form of flu is not fatal, but the patient is very weak and dizzy for a few days. The disease spreads by personal contact and once a person is infected, he stays infectious for about 8 days, after which he is immune to the disease. If the disease spreads at a fast rate, the whole exercise may be jeopardized. On the other hand, extra tents must be flown in for a quarantine area which might be an unnecessary expense, since the exercise is finished in a fortnight. The medical staff decides to send the 6 men back to their barracks and to wait until the next morning before a quarantine is imposed. The next morning 6 more men report sick.

(1) Construct a model for the spread of this disease if no quarantine is imposed.

(2) Calculate from this model the percentage of the men who would have contracted the disease after 8 days.

(3) Write down the boundary value problem to determine the percentage of the men who would have contracted the disease at the end of the exercise. (This type of differential equation is known as a *Ricatti's equation* - see [18] page 59, [50] page 66, or [58] page 60.)

(4) Suppose the military exercise had gone on indefinitely. Write a program to calculate the percentage of men who would have contracted the disease in this case.

Project H^*: Experiment with a Falling Ball

In a laboratory a long ruler is placed in a vertical glass tube. A stroboscope which flashes a 1000 times per minute is set up in front of the tube. A ball is released from rest at the top of the tube in a darkened room. As the ball is released, the stroboscope starts flashing and the flight of the ball is recorded by a camera with an open lens. The result is a photo on which the position (relative to the ruler) of the ball at fixed time instants is recorded. Measuring always from the center of the ball, the following data were obtained:

Time t in seconds	Distance h in meters
0.00	0.00
0.09	0.04
0.15	0.10
0.21	0.20
0.27	0.33
0.33	0.50
0.39	0.70
0.45	0.93
0.51	1.19
0.57	1.47
0.63	1.78
0.67	2.10

(1) Assume that the magnitude of the force of air resistance is directly proportional to the speed. Construct a model to obtain h as a function of t.

(2) Assume that the magnitude of the force of air resistance is directly proportional to the square of the speed. As in §2.9, construct a model to obtain h as a function of t.

(3) In each of the above models an unknown parameter appears in the solution. Use the datapoint at $t = 0.51$ to calculate these parameters.

(4) Write a program for the calculation of these parameters, and which will also plot the data points in the table above as well as each of the graphs of the two models.

(5) What can you say about the validation of these models? Is one model better for implementation than the other?

Project I^*: Altitude of a Rocket

Before attempting this project, first do Exercise 31 in §2.12. The vertical motion of a rocket with linear air resistance was discussed in §2.11. The important output of this model is the position and velocity at any time t during the flight, in particular the maximum altitude and total time for the flight. Write a program which incorporates the following features:

(1) Immediately after the input of the relevant parameters, the program calculates

(a) whether lift-off will in fact occur,

(b) the maximum altitude,

(c) the time for the ascent,

(d) the time for the descent.

(2) The altitude at any time t during the flight (note that there are three stages: ascent with fuel, ascent without fuel, descent).

(3) The velocity at any time t during the flight.

(4) Graphs of the altitude and velocity at any time t.

(5) In the notation of §2.11, use the following parameters in a trial run for your program:

$$\begin{aligned} M &= 25{,}000 \text{ kilograms} \\ n &= 0.8 \\ \alpha &= 400 \text{ meters per second} \\ \beta &= 1{,}000 \text{ kilograms per second} \\ k &= 0.01 \text{ per second} \\ g &= 9.81 \text{ meters per second}^2 \end{aligned}$$

2.14 Mathematical Background

In this section the general theorems used in this chapter are proved. We also discuss some facts about hyperbolic functions, alternating series, and Maclaurin series.

Hyperbolic Functions

We summarize the main results of these convenient functions. For more details see, for example, [16] p. 228. The definitions of these functions are

$$\left.\begin{aligned} \sinh x &= \frac{e^x - e^{-x}}{2} \\ \cosh x &= \frac{e^x + e^{-x}}{2} \\ \tanh x &= \frac{\sinh x}{\cosh x} = \frac{e^x - e^{-x}}{e^x + e^{-x}} = \frac{e^{2x} - 1}{e^{2x} + 1} \end{aligned}\right\} \quad (2.14.1)$$

Just as in trigonometry, one could also define the reciprocals cosech x, sech x, and coth x and the inverse functions arcsinh x, etc. Some of the properties of the hyperbolic functions are:

$$\left.\begin{aligned} \frac{d}{dx}(\sinh x) &= \cosh x \\ \frac{d}{dx}(\cosh x) &= \sinh x \end{aligned}\right\} \qquad (2.14.2)$$

$$\left.\begin{aligned} \cosh(a+b) &= \cosh a \cosh b + \sinh a \sinh b \\ \sinh(a+b) &= \sinh a \cosh b + \cosh a \sinh b \end{aligned}\right\} \qquad (2.14.3)$$

$$(\cosh a)^2 - (\sinh a)^2 = 1 \qquad (2.14.4)$$

The proofs of (2.14.2), (2.14.3), and (2.14.4) are straightforward manipulations of the definitions and will not be given here.

Alternating and Monotonic Decreasing Series

Let $a_1, a_2, a_3, \ldots, a_n, \ldots$ be a sequence of positive real numbers. Then the series

$$a_1 - a_2 + a_3 - a_4 + \ldots + (-1)^{n+1} a_n + \ldots \qquad (2.14.5)$$

is called an *alternating series* because successive terms have opposite signs.

If the sequence has the property that

$$a_{n+1} < a_n$$

for $n = 1, 2, 3, \ldots$ then we say that the series (2.14.5) is *monotonic decreasing*.

Theorem 2.14.1

If $a_1, a_2, a_3, \ldots$ is a monotonic decreasing sequence of positive real numbers with limit zero, then the alternating series (2.14.5) converges. If S denotes the sum of the series and s_n its n-th partial sum, then

$$0 < (-1)^n (S - s_n) < a_{n+1} \qquad (2.14.6)$$

for $n = 1, 2, 3, \ldots$

The proof of this important theorem can be found in [23] page 439. The inequality (2.14.6) is very useful to estimate the error when the sum S is approximated by any partial sum s_n. In particular the case $n = 1$ in (2.14.6) implies

$$a_1 - a_2 < S < a_1 \tag{2.14.7}$$

which gives a simple (but very effective) estimate of the sum S.

Taylor and Maclaurin Series

The *Taylor series* of a function $f(x)$ at the point $x = a$ is defined as

$$f(x) = f(a) + \frac{f^{(1)}(a)}{1!}(x-a) + \frac{f^{(2)}(a)}{2!}(x-a)^2 + \ldots + \frac{f^{(n)}(a)}{n!}(x-a)^n + \ldots \tag{2.14.8}$$

where $f^{(n)}$ denotes the nth derivative of $f(x)$ with respect to x. Given a function $f(x)$ which can be differentiated as many times as we please, the series in (2.14.8) is first determined, then the *interval of convergence* of the series (that is, the values of x for which the series converges) is established, and finally the subinterval on which the equality sign holds is obtained. This interval on which (2.14.8) holds will obviously depend on the function $f(x)$. When $a = 0$ the series is called a *Maclaurin series*. For convenience the following Maclaurin series are given as a reference. For more details see, for example, [16] page 453. The interval in brackets denotes the values of x for which (2.14.8) holds in each case.

$$\begin{aligned}
e^x &= 1 + \frac{x}{1!} + \frac{x^2}{2!} + \frac{x^3}{3!} + \ldots + \frac{x^n}{n!} + \ldots \qquad (-\infty < x < \infty) \\
\sin x &= x - \frac{x^3}{3!} + \frac{x^5}{5!} - \ldots + (-1)^n \frac{x^{2n+1}}{(2n+1)!} + \ldots \quad (-\infty < x < \infty) \\
\sinh x &= x + \frac{x^3}{3!} + \frac{x^5}{5!} + \ldots + \frac{x^{2n+1}}{(2n+1)!} + \ldots \quad (-\infty < x < \infty) \\
\cos x &= 1 - \frac{x^2}{2!} + \frac{x^4}{4!} + \ldots + (-1)^n \frac{x^{2n}}{(2n)!} + \ldots \quad (-\infty < x < \infty) \\
\cosh x &= 1 + \frac{x^2}{2!} + \frac{x^4}{4!} + \ldots + \frac{x^{2n}}{(2n)!} + \ldots \qquad (-\infty < x < \infty) \\
\ln(1+x) &= x - \frac{x^2}{2} + \frac{x^3}{3} + \ldots + (-1)^n \frac{x^{n+1}}{n+1} + \ldots \quad (-1 < x \le 1) \\
\arctan x &= x - \frac{x^3}{3} + \frac{x^5}{5} - \ldots + (-1)^n \frac{x^{2n+1}}{2n+1} + \ldots \quad (-1 \le x \le 1)
\end{aligned}$$

$$\begin{aligned}
\text{arctanh } x &= x + \frac{x^3}{3} + \frac{x^5}{5} + \ldots + \frac{x^{2n+1}}{2n+1} + \ldots \qquad (-1 < x < 1) \\
(1+x)^m &= 1 + \frac{m}{1!}x + \frac{m(m-1)}{2!}x^2 + \ldots \\
&\quad + \frac{m(m-1)\ldots(m-n+1)}{n!}x^n + \ldots \\
&\quad (\text{where } m \text{ can be any real number and } -1 < x < 1)
\end{aligned}$$

Proof of Theorem 2.4.1

To prove Theorem 2.4.1 we first need a lemma called Gronwall's lemma:

Lemma

If $u(t)$ and $g(t)$ are continuous functions, and

$$u(t) \leq c + \int_a^t g(\tau)u(\tau)\, d\tau \tag{2.14.9}$$

for all t in the interval $a \leq t \leq T$, with c a non-negative constant and $g(t) \geq 0$ on this interval, then

$$u(t) \leq c e^{\int_a^t g(\tau)\, d\tau} \tag{2.14.10}$$

Proof

Let $h(t)$ denote the right hand side of (2.14.9). Then $h(t)$ is differentiable on the open interval (a, T), $h(a) = c$ and

$$\frac{dh}{dt} = g(t)u(t)$$

(See [16] page 185.) Rewrite (2.14.9) in terms of h by multiplying the inequality by $g(t)$:

$$\frac{dh}{dt} \leq g(t)h(t)$$

(Note that $g(t) \geq 0$!). Now multiply this inequality by the usual integrating factor (2.1.13) for linear differential equations and use the product rule for differentiation to obtain

$$\frac{d}{dt}\left[h(t)e^{\int_a^t g(\tau)\,d\tau}\right] \le 0$$

Since the slope of the function in brackets is always non-negative, and the function is continuous, it follows that

$$h(t)e^{-\int_a^t g(\tau)\,d\tau} \le h(a)e^0$$

$$h(t) \le ce^{\int_a^t g(\tau)\,d\tau}$$

But h(t) is the right hand side of (2.14.9), and hence (2.14.10) follows.

Theorem 2.4.1

Let $f(t,y)$ be continuous on the rectangle $R = \{(t,y) : a \le t \le b, c \le y \le d\}$ and let $f(t,y)$ satisfy a Lipschitz condition in y on $[c,d]$ with the same constant k for every t in $[a,b]$. If a solution of the boundary value problem

$$\begin{aligned} \frac{dy}{dt} &= f(t,y) \\ y(a) &= \alpha \end{aligned} \qquad (2.14.11)$$

exists in R, then the solution is unique and the solution is a continuous function of the boundary value α in R.

Proof

Let $y(t)$ and $z(t)$ be any two solutions of (2.14.11). In integral form we have then:

$$\begin{aligned} y(t) &= \alpha + \int_a^t f(\tau,y)\,d\tau \\ z(t) &= \alpha + \int_a^t f(\tau,z)\,d\tau \end{aligned}$$

Subtract the equations and take absolute values:

$$|y(t) - z(t)| = \left|\int_a^t [f(\tau,y) - f(\tau,z)]\,d\tau\right|$$

$$\leq \int_a^t |f(\tau,y) - f(\tau,z)| \, d\tau$$
$$\leq \int_a^t k|y-z| \, d\tau$$

Now apply Gronwall's lemma with $u = |y-z|$, $c = 0$, and $g = k$ in (2.14.9), then (2.14.10) implies that $|y(t) - z(t)| \leq 0$ for all t in the closed interval $[a, b]$. Hence, $y(t) = z(t)$ for $a \leq t \leq b$, which proves the uniqueness of the solution.

Let $x(t)$ be a solution of the boundary value problem

$$\begin{cases} \frac{dx}{dt} = f(x,t) \\ x(a) = \beta \end{cases}$$

in R, then again by integrating, subtracting, and taking absolute values, it follows that

$$\begin{aligned} |y(t) - x(t)| &\leq |\alpha - \beta| + \left| \int_a^t [f(\tau,y) - f(\tau,x)] \, d\tau \right| \\ &\leq |\alpha - \beta| + \int_a^t k|y-x| \, d\tau \end{aligned}$$

Apply Gronwall's lemma with $u = |y-x|$, $c = |\alpha - \beta|$, and $g = k$ in (2.14.9), then (2.14.10) implies that

$$|y(t) - x(t)| \leq |\alpha - \beta| e^{k(t-a)} \leq |\alpha - \beta| e^{k(b-a)}$$

since $t \leq b$ and $k > 0$. Hence $|y-x|$ can be made arbitrarily small by making $|\alpha - \beta|$ small enough. This completes the proof.

3

Numerical Methods

3.1 Introduction

In Chapter 2 several different types of differential equations emerged as a result of the modelling process. Fortunately these equations could be solved explicitly. However, as was mentioned in §1.2, this is not always possible. Consider, for example, the mathematical model constructed in §2.11 for a rocket being launched vertically. Instead of air resistance proportional to the speed, let us make the more realistic assumption that the air resistance is proportional to some exponent n of the speed. Then we obtain the initial value problem

$$\left.\begin{aligned} &\frac{dv}{dt} + \frac{\beta k}{M - \beta t} v^n = \frac{\alpha\beta}{M - \beta t} - g \\ &v(0) = 0 \end{aligned}\right\} \tag{3.1.1}$$

which is nonlinear in the speed v. If we also assume that the rocket rises to an altitude where the gravitational acceleration g is not constant, the differential equation becomes even more complicated. Since the solution cannot be obtained explicitly, the only other alternative is to calculate the solution approximately at a finite number of points. For this we need some appropriate numerical algorithm.

In this chapter we shall look at different methods to find an approximation of the solution of a boundary value problem. We had already seen in §1.3 that some idea of the solution curve can be obtained by drawing the direction field of the equation

$$\frac{dy}{dx} = F(x, y), \qquad y(a) = \alpha \tag{3.1.2}$$

We shall now start at the initial value $(a, y(a))$ and construct from there an approximate solution curve by connecting certain points in the (x, y)-plane with straight lines. The points in question will be calculated by some appropriate numerical algorithm.

In this chapter we shall only consider first order differential equations of the type (3.1.2). It must be understood that the algorithms used in computer programs are very sophisticated and complicated. The aim in this chapter is to convey the flavour and the basics of numerical algorithms for first order differential equations, and certainly not to develop the most efficient algorithm, or even to give an overview of the different types of algorithms. One reason for this viewpoint is that in practice you will probably select an efficient computer package on the market at that time, rather than write your own algorithm. Another reason is that the construction of new algorithms and the improvement of known algorithms require a thorough understanding of approximation theory, which is a field of study on its own. Should you be interested in this field, then this chapter will give you a brief introduction.

One important warning: do not plunge immediately into some numerical algorithm when you realize that the problem cannot be solved explicitly - it may be more efficient to transform the problem, keeping a close watch on possible singularities and/or instability. Remember that a computer dishes out numbers when you load an algorithm with appropriate starting values. The point is whether the output means anything at all, and if so, whether the output is indeed an approximation of the solution that you are seeking.

3.2 Existence Theorem

When a solution of a boundary value problem cannot be found explicitly, the obvious question is whether a solution does indeed exist. Should the answer be negative, it would be futile to implement some numerical algorithm, because the numerical values produced by the algorithm would be meaningless. Equally important is the question of uniqueness; otherwise we shall not know to which solution (or even perhaps a mixture of the solutions!) the calculated approximate values refer. (We shall give an example in the reading material at the end of the section.) It is, therefore, crucially important to settle the question of existence and uniqueness of the solution before an algorithm is implemented. There are many different existence theorems for different boundary value problems in the mathematical literature. For our purposes the following theorem is appropriate - compare it with Theorem 2.4.1 in §2.4.

Theorem 3.2.1

Let $f(t,y)$ be continuous on the rectangle $R = \{(t,y) : |t-t_0| \leq a, |y-y_0| \leq b\}$ and let $f(t,y)$ satisfy a Lipschitz condition in y on the closed interval $[y_0 - b, y_0 + b]$ for each t in the closed interval $[t_0 - a, t_0 + a]$. Then there exists a unique solution of the boundary value problem

$$\frac{dy}{dt} = f(t,y), \qquad y(t_0) = y_0 \tag{3.2.1}$$

on the interval $I_\delta = \{t : |t - t_0| \leq \delta\}$ where $\delta =$ minimum $\{a, \frac{b}{M}\}$ with M the maximum of $|f(t,y)|$ on R.

The proof of this very important theorem can be found in §3.9 where only the existence is proved, since the uniqueness was settled in §2.14 beforehand. To illustrate the use of Theorem 3.2.1, consider the nonlinear initial value problem

$$\left.\begin{aligned} &\frac{dy}{dt} = 1 + y^2, \\ &y(0) = 0 \end{aligned}\right\} \tag{3.2.2}$$

In this case $f(t,y) = 1 + y^2$, and we immediately note that f is continuous on the rectangle $R = \{(t,y) : |t| \leq a, |y| \leq b\}$ with a and b arbitrarily large. Moreover, f has a continuous derivative $2y$ with respect to y, and hence, f also satisfies a Lipschitz condition in y on the interval $\{y : |y| \leq b\}$. Thus, the conditions of Theorem 3.2.1 are satisfied, and so there exists a unique solution on the interval $\{t : |t| \leq \delta\})$ where

$$\delta = \frac{b}{1+b^2} \qquad (b \text{ arbitrarily large})$$

If we sketch the graph of δ as a function of b, it is clear that the largest value for δ is 0.5 when $b = 1$. So Theorem 3.2.1 guarantees that there exists a unique solution on the interval $\{t : |t| \leq 0.5\}$. Moreover, on this interval the absolute value of this solution will never exceed 1.

We chose this example because we can, in fact, find the solution explicitly. This enables us to compare the solution with the information given by Theorem 3.2.1. Elementary integration shows that $y = \tan t$ is the solution of (3.2.2). Note that this solution exists on the interval $|t| < \frac{\pi}{2}$

so that Theorem 3.2.1 does not provide the largest interval where the solution exists. However, if we take the bound $y \leq 1$ into consideration, then the solution shows that $|t| \leq \frac{\pi}{4}$, which should be compared to the interval $|t| \leq 0.5$ given by the theorem.

- Read: The short paper [52] on the importance of the existence and uniqueness of a solution.
- Do: Exercises 1, 2 , 3 in §3.7.

3.3 Euler Algorithm

The granddaddy of a whole family of algorithms was named after the greatest mathematician in the history of Switzerland, Léonard Euler (1707 – 1783). It is estimated that if all the mathematical papers by Euler were collected in one publication, it would consist of 80 hefty volumes!

- Read: A short biography on Euler in [5], pages 151 to 165.

The Euler algorithm is certainly the simplest method to obtain an approximate solution to the boundary value problem (3.2.1). Let y_i^* denote the approximate value of the solution y at $t = t_i$. Then the Euler algorithm is the following recursive procedure for $i = 0, 1, 2, \ldots$:

$$\left.\begin{aligned} y_0^* &= y_0 \\ y_{i+1}^* &= y_i^* + hf(t_i, y_i^*) \end{aligned}\right\} \qquad (3.3.1)$$

where h is a given step length and $t_{i+1} = t_i + h$, recursively.

Graphically the algorithm means that a tangent to the curve $y(t)$ is drawn at (t_0, y_0), and that y_1^* is the $y-$value at $t = t + h = t_1$ on this tangent. If h is small, then y_1^* is a reasonable approximation to $y(t_1)$, as seen in Figure 3.3.1 below. At (t_1, y_1^*) a line is drawn with the same slope as the tangent to a solution curve passing through this point (in other words, we draw an element of the direction field at this point - see §1.3), and y_2^* is the y-value at $t = t + 2h = t_2$ on this line. By repeating this process at (t_2, y_2^*), a sequence of points (t_i, y_i^*) is generated recursively for $i = 1, 2, 3, \ldots$. If we connect these points (t_i, y_i^*) with straight line segments, we obtain an approximate function $y^*(t)$ of the solution $y(t)$ as shown in Figure 3.3.1.

- Do: Exercises 4, 5 in §3.7.

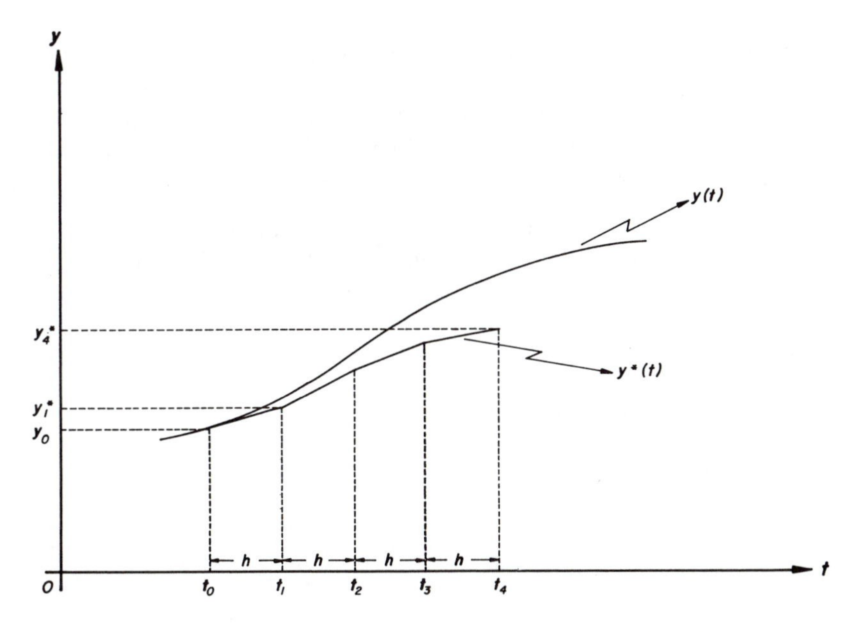

Figure 3.3.1: **Approximation by the Euler algorithm**

It is interesting to note that the Euler algorithm can be seen in three different ways:

(1) If the derivative in (3.2.1) at the point $t = t_i$ is approximated by the expression

$$\frac{y(t_{i+1}) - y(t_i)}{h}$$

then (3.3.1) follows immediately from (3.2.1).

(2) If (3.2.1) is integrated over the interval $[t_i, t_{i+1}]$, and the integral on the right hand side is approximated by a simple left endpoint rectangle rule (see §3.9)

$$y(t_{i+1}) - y(t_i) = \int_{t_i}^{t_{i+1}} f(t, y(t))\, dt \approx hf(t_i, y(t_i)) \qquad (3.3.2)$$

then (3.3.1) follows immediately.

(3) If we assume that the solution $y(t)$ of (3.2.1) can be written as a Taylor series at the point $t = t_i$ (see (2.14.8) in §2.14), we have

$$y(t_{i+1}) = y(t_i) + hf(t_i, y(t_i)) + \frac{h^2}{2} y^{(2)}(t_i) + \frac{h^3}{6} y^{(3)}(t_i) + \ldots \quad (3.3.3)$$

Taking the first two terms on the right hand side as an approximation, the algorithm (3.3.1) results.

Each of these three viewpoints shows an obvious way in which the Euler algorithm can be improved. For example, in (2) we can expect better results if the trapezium rule or Simpson's rule (see §3.9 or [16] page 483) is used to approximate the integral. In §3.5 we shall show another way in which the Euler algorithm can be improved.

Having done Exercises 4 and 5, you will agree that we need a computer program to do the hard work! Let us first write down the algorithm:

Algorithm: EULER

(1) Specify t_0, y_0, h, and n (n is an integer).

(2) Define the function $f(t, y)$.

(3) Print (t_0, y_0).

(4) Let $i = 0$.

(5) Calculate $y_{i+1} = y_i + hf(t_i, y_i)$.

(6) Let $t_{i+1} = t_i + h$.

(7) Print (t_{i+1}, y_{i+1}).

(8) If $i + 1 = n$ then stop; otherwise continue to the next step.

(9) Let $i = i + 1$.

(10) Go to step 5

This algorithm must be translated into a program for the computer, according to the computer language of the reader's choice. Consider the problem

$$\frac{dy}{dt} + y = t + 1, \qquad y(0) = 1.$$

In TURBO PASCAL 4.0 a program would look like this:

```
program EULER;
uses
        Printer;
var
        i,n : integer;
        h,t,y : real;
function f(t,y: real): real;
begin
        f:=-y+t+1;
end;
```

```
begin
    writeln('Number of iterations =');readln(n);
    writeln('Step length =');readln(h);
    writeln('Initial value of t =');readln(t);
    writeln('Initial value of y =');readln(y);
    writeln(Lst,' t ',' y ');
    for i:=1 to n do
    begin
        writeln(Lst,t:7:3,y:7:3);
        y:=y+h*f(t,y);
        t:=t+h;
    end;
end.
```

This program will print a table of values which represents points on the approximate curve. One could also combine this program and the program OURGRAPH in Chapter 1 to obtain a graph of the approximate curve.

- Do: Exercises 6, 7, 8 in §3.7.
- Do: Project A in §3.8.

3.4 Error Analysis

The approximate solution $y^*(t)$ shown in Figure 3.3.1 depends on the step length h. One would expect that $y^*(t)$ will tend to the solution $y(t)$ as h tends to zero. If this does happen, we say that the Euler algorithm *converges*. It is also important to know the *error* of the approximation for a given h. We shall now analyse this error and show how the error can be estimated.

It can be shown that the Euler algorithm always converges, provided that the conditions of Theorem 3.2.1 are satisfied. (See, for example, [32] page 15 where the Euler algorithm is used to prove a theorem like Theorem 3.2.1.) Unfortunately, the convergence is slow, meaning that h must be made very small to obtain the approximate solution within a specified accuracy. Obviously, the smaller h is, the more steps (or points t_i) there are in a fixed interval, and hence more calculations must be performed. Thus slow convergence means more work, and for this reason the Euler algorithm is not very popular when a high accuracy is important.

If we assume that the initial data y_0 is accurate, then there are two types of errors which normally occur in any numerical algorithm. They are *round-off error* and *truncation error*. We shall define these errors precisely, and then show how the error for the Euler algorithm can be estimated.

In any calculation which involves irrational numbers, we automatically approximate every irrational number by a rational number because we cannot handle an infinite number of decimal digits. For example, we write the rational number 1.4142136 for the irrational number $\sqrt{2}$. Thus, instead of the value y_i^* (given by the algorithm) we, in fact, write down the rational number y_i^{**} which is y_i^* rounded off to a predetermined number of decimal places (which is of course dependent on the accuracy built into the calculator being used). In general, the round-off error is the difference $y_i^* - y_i^{**}$. In the case of the Euler algorithm it must be noted that in the calculation of y_i^* the value y_{i-1}^{**} is used, by (3.3.1). Hence the following definition makes sense:

Definition 3.1

The round-off error r_i at the i-th step $(i = 1, 2, \ldots)$ of the Euler algorithm is

$$r_i = y_{i-1}^{**} + hf(t_{i-1}, y_{i-1}^{**}) - y_i^{**} \qquad (3.4.1)$$

It is rather difficult to analyse the round-off error because it is determined by the number of decimal places in the calculation, the type of calculator being used, the kind of arithmetical operations involved and their sequence in the algorithm (see Exercise 9 in §3.7), etc. Usually a bound is determined for the round-off error at each step. In practice it is sometimes easier to repeat the algorithm, using an accuracy of more decimal places to see if the answer is different. It at least gives an indication whether round-off errors are important in the calculation.

- Do: Exercises 9, 10 in §3.7.

Since the algorithm produces a value y_i^* which approximates the value $y_i = y(t_i)$ of the solution, an error inherent in the algorithm is also produced. This *truncation error* is the difference between y_i and the value produced at $t = t_i$ when the exact value of the solution y_{i-l} is used in the algorithm.

Definition 3.2

The truncation error T_i at the i-th step ($i = 1, 2, 3, \ldots$) *of the Euler algorithm is*

$$T_i = y_i - y_{i-1} - hf(t_{i-1}, y_{i-1}) \tag{3.4.2}$$

Note that the truncation error is the amount erred at each step, and not the accumulated error $y_i - y_i^*$ (except when $i = 1$), as shown in Figure 3.4.1. If $f(t, y)$ is smooth enough, the truncation error can be analysed in the following manner. Assume that $y(t)$ can be written

$$y(t_i) = y(t_{i-1}) + (t_i - t_{i-1})y^{(1)}(t_{i-1}) + \tfrac{1}{2}(t_i - t_{i-1})^2 y^{(2)}(\tau_i)$$

where the exponents in brackets denote derivatives with respect to t as in §2.14, and where $t_{i-1} \leq \tau_i \leq t_i$ (see [16] page 449). By (3.4.2) and (3.2.1) it immediately follows that

$$T_i = \frac{h^2}{2} y^{(2)}(\tau_i) \tag{3.4.3}$$

Hence the truncation error is proportional to the square of the step length, provided that $f(t, y)$ is smooth enough.

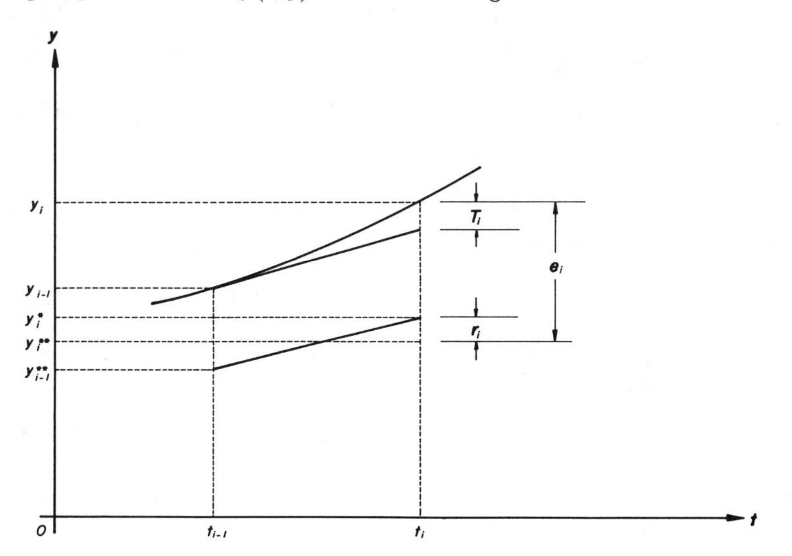

Figure 3.4.1: **Relation between the different error types**

We can now combine the round-off and truncation errors to define and analyse the total error of the algorithm.

Definition 3.3

The error e_i at the i–th step $(i = 1, 2, \ldots)$ of the Euler algorithm is

$$e_i = y_i - y_i^{**} \tag{3.4.4}$$

For convenience introduce

$$G_i = \begin{cases} 0 & \text{if } y = y_i^{**} \\ \dfrac{f(t_i, y_i) - f(t_i, y_i^{**})}{y_i - y_i^{**}} & \text{if } y_i \neq y_i^{**} \end{cases} \tag{3.4.5}$$

and note that $f(t, y)$ is assumed to satisfy a Lipschitz condition in y for all t (by Theorem 3.2.1):

$$|f(t, y_i) - f(t, y_i^{**})| \leq G|y_i - y_i^{**}|$$

Hence

$$|G_i| \leq G \quad \text{for} \quad i = 1, 2, \ldots \tag{3.4.6}$$

Now add equations (3.4.1) and (3.4.2) to obtain

$$\begin{aligned} r_i + T_i &= y_i - y_i^{**} - (y_{i-1} - y_{i-1}^{**}) - h[f(t_{i-1}, y_{i-1}) - f(t_{i-1}, y_{i-1}^{**})] \\ &= e_i - e_{i-1} - hG_{i-1}e_{i-1} \end{aligned}$$

by (3.4.4) and (3.4.5). This provides an equation which expresses the error e_i at the i–th step in terms of the error at the $(i-1)$–th step:

$$e_i = (1 + hG_{i-1})e_{i-1} + r_i + T_i \tag{3.4.7}$$

The values r_i and T_i are generally not known, but as in (3.4.6) they can usually be estimated in the form

$$|r_i| \leq r \quad \text{and} \quad |T_i| \leq T \tag{3.4.8}$$

where r and T are independent of i, but could be a function of h. By taking absolute values on both sides of (3.4.7) we find

$$|e_i| \leq (1 + hG)|e_{i-1}| + r + T \qquad (i = 1, 2, \ldots) \tag{3.4.9}$$

The inequality provides a recursive relation by which e can be found. Substitute the values $i = 1, 2, \ldots$

$$\begin{aligned}
|e_1| &\leq (1+hG)|e_0| + r + T \\
|e_2| &\leq (1+hG)|e_1| + r + T \\
&\leq (1+hG)^2|e_0| + (1+hG)(r+T) + r + T \\
|e_3| &\leq (1+hG)|e_2| + r + T \\
&\leq (1+hG)^3|e_0| + [(1+hG)^2 + (1+hG) + 1](r+T)
\end{aligned}$$

Note that the expression in square brackets is the geometric series

$$1 + v + v^2 + \ldots + v^{n-1} = \frac{v^n - 1}{v - 1} \qquad (v \neq 1)$$

By induction it follows that

$$|e_i| \leq (1+hG)^i|e_0| + \frac{(1+hG)^i - 1}{hG}(r+T) \tag{3.4.10}$$

If the initial value is accurate, we have $e_0 = 0$, and the following error estimate results

$$|e_i| \leq \frac{r+T}{hG}[(1+hG)^i - 1]$$

Another popular estimate immediately follows if the inequality

$$(1+x)^i \leq e^{ix}$$

is utilized, and $ih = t - t_0$ is substituted:

$$|e_i| \leq \frac{r+T}{hG}[e^{G(t_i - t_0)} - 1] \qquad (i = 1, 2, \ldots) \tag{3.4.11}$$

We can now estimate the error at each step of the Euler algorithm, provided that the round-off and truncation errors can be estimated as in (3.4.8).

We know that the Euler algorithm converges if the function $f(t, y)$ satisfies the conditions of Theorem 3.2.1. Hence if h is made smaller, the error must decrease. There is, however, one practical aspect of importance. Suppose that f is smooth enough so that (3.4.3) can be used in the form

$$|T_i| \leq Kh^2$$

where K is some constant. Then (3.4.11) contains the expression

$$\frac{r + Kh^2}{hG}.$$

The graph of this function shows that the function decreases as h decreases until $h = \sqrt{r/K}$. If h decreases further, the function (and hence also the error estimate) will *increase*, and, in fact, grow arbitrarily large as h tends to zero. Hence an arbitrarily small h will not necessarily produce an arbitrarily accurate solution, due to effect of a non-zero round-off error.

Example

Consider the boundary value problem

$$\left.\begin{aligned} &\frac{dy}{dt} = \frac{1}{y} \qquad (0 < t < \infty) \\ &y(0) = 1 \end{aligned}\right\} \tag{3.4.12}$$

By elementary integration the solution is $y = \sqrt{2t+1}$. This enables us to see what the exact error of the Euler algorithm is, so that the predicted error by (3.4.11) can be compared.

Suppose we need the value $y(1)$ accurate to two decimal places. The question is how small h must be chosen to achieve this. We shall use (3.4.11) to answer this question, and so must first determine r, T, and G.

Due to the accuracy of the calculator used (12 decimal places) in comparison with the accuracy required, we can take $r = 0$. To determine T we utilize (3.4.3), which requires an estimate of the second derivative of y. By (3.4.12) it follows that

$$\frac{d^2y}{dt^2} = -\frac{1}{y^2}\frac{dy}{dt} = -\frac{1}{y^3}.$$

Moreover, the slope of y is always positive, since the initial value of y is positive, and so the right hand side of (3.4.12) will always be positive. Hence y is an increasing function of t so that $y(t) \geq 1$ for all $t \geq 0$. Now we obtain by (3.4.3) that

$$|T_i| = \frac{1}{2}h^2|y^{-3}(\tau_i)| \leq \frac{1}{2}h^2$$

so that we can take $T = \frac{h^2}{2}$ in this case. To find the constant G of the Lipschitz condition (see (3.4.6)), we may use the partial derivative with respect to y of $f(t,y) = y^{-1}$ as in §2.4, since this function f has a continuous derivative:

$$|\frac{\partial f}{\partial y}| = |-y^{-2}| \leq 1$$

Hence we take $G = 1$. Substitute these values for r, T, and G in (3.4.11) to obtain an estimate of the error:

$$|e_i| \leq \frac{1}{2}h[e^{t_i} - 1] \qquad (3.4.13)$$

When $t_i = 1$, we find that $|e_i| \leq 0.86h$. To ensure accuracy to two decimal places, we need $|e_i| < 0.001$, which means that $h < 0.0012$ by (3.4.13). If we take $h = 0.001$, we must apply the Euler algorithm a *thousand* times to be sure that the value of $y(1)$ is correct to *two* decimal places. This is really slow convergence! Of course this value of h may be due to crude estimates, and that the convergence is, in actual fact, not so slow. To investigate this, we show in Table 3.4.1 the exact error for two different values of h.

t_i	$y_i = \sqrt{2t_i + 1}$	y_i^{**} for h_1	y_i^{**} for h_2	$y_i - y_i^{**}$ for h_1	$y_i - y_i^{**}$ for h_2
0.05	1.049	1.050	1.049	-0.001	0.000
0.10	1.095	1.098	1.096	-0.002	0.000
0.15	1.140	1.143	1.140	-0.003	0.000
0.20	1.183	1.187	1.184	-0.004	0.000
0.25	1.225	1.229	1.225	-0.004	0.000
0.30	1.265	1.270	1.265	-0.005	0.000
0.35	1.304	1.309	1.304	-0.005	-0.001
0.40	1.342	1.347	1.342	-0.006	-0.001
0.45	1.378	1.384	1.379	-0.006	-0.001
0.50	1.414	1.421	1.415	-0.006	-0.001
0.55	1.449	1.456	1.450	-0.007	-0.001
0.60	1.483	1.490	1.484	-0.007	-0.001
0.65	1.517	1.524	1.517	-0.007	-0.001
0.70	1.549	1.556	1.550	-0.007	-0.001
0.75	1.581	1.589	1.582	-0.007	-0.001
0.80	1.612	1.620	1.613	-0.008	-0.001
0.85	1.643	1.651	1.644	-0.008	-0.001
0.90	1.673	1.681	1.674	-0.008	-0.001
0.95	1.703	1.711	1.704	-0.008	-0.001
1.00	1.732	1.740	1.733	-0.008	-0.001

Table 3.4.1: **The Euler algorithm applied to (3.4.12) with $h_1 = 0.5$ and $h_2 = 0.005$**

Table 3.4.1 shows that the convergence is indeed slow; in fact $h = 0.05$ is not small enough to ensure accuracy to two decimal places. On the other hand, $h = 0.005$, or equivalently 200 iterations, does provide accuracy to two decimal places, which is better than the 1,000 iterations given by the estimate (3.4.13).

- Do: Exercises 11, 12, 13, 14 in §3.7.

3.5 Runge-Kutta Algorithm

There are many improvements of the Euler algorithm. One type is the so-called Runge-Kutta algorithms which came about originally in the period 1895 - 1901 due to the research of C. Runge, W. Kutta, and K. Heun, with further improvements since then. In this section we shall briefly look at these Runge-Kutta algorithms in relation to the Euler algorithm. In (3.3.3) it was shown that the Euler algorithm can be seen as the first two terms in a Taylor series. If we take more terms in the series, we expect to improve the accuracy, but at the cost of the calculation of the second and higher derivatives of $y(t)$ at each step. The calculation of these derivatives can involve a lot of work, especially when f is a complicated function because

$$\frac{d^2y}{dt^2} = \frac{\partial f}{\partial t} + \frac{\partial f}{\partial y} f \tag{3.5.1}$$

where f refers to the boundary value problem (3.2.1). The merit of the Runge-Kutta algorithms is that it produces the same accuracy obtained when the first three terms in (3.3.3) are taken, but without the necessity of calculating the second derivative. To show this we shall first look at a more general set-up.

Any algorithm of the form

$$y_{i+1}^* = y_i^* + h\phi(t_i, y_i^*, h) \tag{3.5.2}$$

where the approximation y_{i+1}^* is calculated only with the aid of h, t_i, and y_i^*, is called a *one-step method*. For example, in the Euler algorithm we have $\phi(t_i, y_i^*, h) = f(t_i, y_i^*)$ by (3.3.1). There the function ϕ represents the slope of the line segments in the approximation.

We can generalize the Euler algorithm in the following way: take one or more points in the closed interval $[t_i, t_{i+1}]$, calculate the slope $f(t, y)$ at each point, and then take ϕ as an average of these slopes. Consider the case of only one point $t_i + \alpha h$ in the interval $[t_i, t_{i+1}]$, where α is a real number between zero and one which is as yet undetermined.

Let η_α and $\eta_{\alpha+1}$, respectively, denote the y-values at $t + \alpha h$ and t_{i+1} obtained when the Euler algorithm is used on the intervals $[t_i, t_i + \alpha h]$, and $[t_i + \alpha h, t_{i+1}]$, consecutively. We have by (3.3.1)

$$\left.\begin{aligned} \eta_\alpha &= y_i^* + \alpha h f(t_i, y_i^*) \\ \eta_{i+1} &= \eta_\alpha + (1-\alpha) h f(t_i + \alpha h, \eta_\alpha) \end{aligned}\right\} \tag{3.5.3}$$

and so, instead of y_{i+1}^* in (3.3.1) a new algorithm is obtained by combining the two equations of (3.5.3):

$$\eta_{i+1} = y_i^* + h[\alpha f(t_i, y_i^*) + (1-\alpha) f(t_i + \alpha h, \eta_\alpha)] \tag{3.5.4}$$

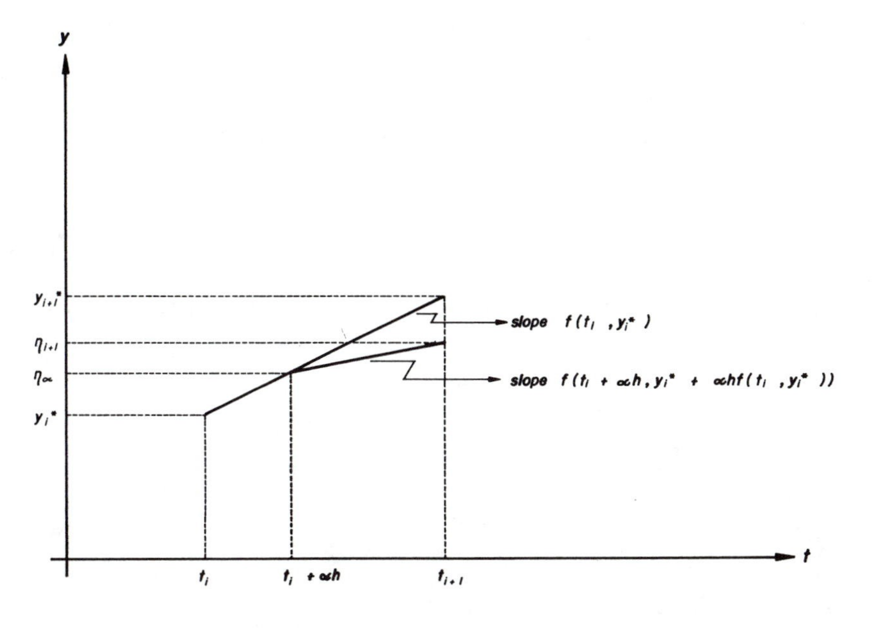

Figure 3.5.1: **Construction of function ϕ as an average of slopes**

When we compare (3.5.2) and (3.5.4), we see that ϕ is the expression in square brackets, which is just a linear combination of the slopes $f(t_i, y_i^*)$ and $f(t_i + \alpha h, \eta_\alpha)$.

Consider then the more general linear combination

$$\phi(t_i, y_i^*, h) = af(t_i, y_i^*) + bf(t_i + \alpha h, y_i^* + \alpha h f(t_i, y_i^*)) \qquad (3.5.5)$$

where $0 < \alpha \leq 1$ and a, b, and α must be determined to give the best possible accuracy. As we noted at the beginning of this section, we are interested in the first three terms of the Taylor series. Let us require in (3.5.2) that if both sides are written as Taylor series, then the first three terms of the two series must be identical.

The Taylor expansion of the left hand side of (3.5.2) is

$$y_i^* + hf(t_i, y_i^*) + \tfrac{1}{2}h^2[f_t(t_i, y_i^*) + f_y(t_i, y_i^*)f(t_i, y_i^*)] + \ldots$$

where the subscripts y and t denote partial derivatives and where the differential Equations (3.2.1) and (3.5.1) were used for the first and second derivatives, respectively.

The Taylor expansion of the function of two variables $f(t_i + \alpha h, y_i^* + \alpha h f(t_i, y_i^*))$ on the right hand side of (3.5.2) is (see [64] page 45 or [57] page 157)

$$y_i^* + haf(t_i, y_1^*) + hb[f(t_i, y_i^*) + \alpha h f_t(t_i, y_i^*) + \alpha h f(t_i, y_i^*) f_y(t_i, y_i^*)] + \dots$$
$$= y_i^* + (a+b)hf(t_i, y_i^*) + \alpha b h^2 [f_t(t_i, y_i^*) + f(t_i, y_i^*) f_y(t_i, y_i^*)] + \dots$$

When the terms in h and h^2 are compared, it follows that

$$a + b = 1 \tag{3.5.6}$$
$$\alpha b = \frac{1}{2} \tag{3.5.7}$$

After solving for a and b in terms of α, an improved algorithm follows by (3.5.2) and (3.5.5):

$$y_{i+1}^* = y_i^* + \frac{2\alpha - 1}{2\alpha} k_0 + \frac{1}{2\alpha} k_1 \tag{3.5.8}$$

where α is any real number with $0 < \alpha \le 1$ and k_0 and k_1 are given by

$$\left. \begin{array}{l} k_0 = hf(t_i, y_i^*) \\ k_1 = hf(t_i + \alpha h, y_i^* + \alpha k_0) \end{array} \right\} \tag{3.5.9}$$

The truncation error

$$T_i = y_i - y_{i-l} - h\phi(t_{i-1}, y_{i-1}, h)$$

is analysed in the same manner as in §3.4 (see (3.4.2) for the case of the Euler algorithm). Instead of (3.4.3) we now obtain

$$T_i = \frac{h^3}{12}[3\alpha \frac{d^2 y}{dt^2} \frac{\partial f}{\partial y} + (2\alpha - 3)\frac{d^3 y}{dt^3}] + O(h^4) \tag{3.5.10}$$

where the functions in the square brackets are again calculated at intermediate points in the interval $[t_{i-1}, t_i]$. The important point to note is that the terms in h and h^2 cancelled out due to the above construction of the algorithm, with the result that the truncation error is proportional to the cube of the step length provided, of course, that $f(t, y)$ is smooth enough. The error estimate (3.4.11) is still valid with the same G as in §3.4 (see, for example, [9] page 181).

Two special cases of (3.5.8) are frequently used. The first case when $\alpha = \frac{1}{2}$ is known as the *modified Euler algorithm*:

$$y_{i+1}^* = y_i^* + hf(t_i + \tfrac{1}{2}h, y_i^* + \tfrac{1}{2}hf(t_i, y_i^*)) \tag{3.5.11}$$

Note that this algorithm results when the integral in (3.3.2) is approximated by

$$\int_{t_i}^{t_{i+1}} f(t, y(t))\, dt \approx hf(t_i + \frac{t_{i+1} - t_i}{2}, y(t_i + \frac{t_{i+1} - t_i}{2}))$$

and the Euler algorithm is used to calculate $y(t_i + \frac{1}{2}h)$. Hence, the difference between this algorithm and the Euler algorithm is that the integral is approximated by rectangles with the height of each rectangle taken at the midpoint of the base in this algorithm and the left endpoint of the base in the Euler algorithm.

The other case when $\alpha = 1$ is known as the *Heun algorithm*:

$$y_{i+1}^* = y_i^* + \tfrac{1}{2}h[f(t_i, y_i^*) + f(t_{i+1}, y_i^* + hf(t_i, y_i^*))] \qquad (3.5.12)$$

This algorithm results when the integral in 3.3.2 is approximated by the trapezium rule

$$\int_{t_i}^{t_{i+1}} f(t, y(t))dt \approx \tfrac{1}{2}h[f(t_i, y(t_i)) + f(t_{i+1}, y(t_{i+1}))]$$

where the Euler algorithm is again used to calculate $y(t_{i+1})$.

- Do: Exercises 15, 16 in §3.7.

The choice of ϕ in (3.5.12) came about when one additional point was used in the interval $[t_i, t_{i+1}]$ and a linear combination of the two values of the function f was taken. A linear combination of more values of f should lead to more accurate algorithms. A well-known example is the popular *Runge-Kutta algorithm*:

$$y_{i+1}^* = y_i^* + \tfrac{1}{6}(k_0 + 2k_1 + 2k_2 + k_3) \qquad (3.5.13)$$

with the k_i given by

$$\left.\begin{aligned} k_0 &= hf(t_i, y_i^*) \\ k_1 &= hf(t_i + \tfrac{1}{2}h, y_i^* + \tfrac{1}{2}k_0) \\ k_2 &= hf(t_i + \tfrac{1}{2}h, y_i^* + \tfrac{1}{2}k_1) \\ k_3 &= hf(t_i + h, y_i^* + k_2) \end{aligned}\right\} \qquad (3.5.14)$$

Note that k_0 and k_l in (3.5.9) with $\alpha = \frac{1}{2}$ are the same as k_0 and k_l in (3.5.14). We call the Runge-Kutta algorithm a *fourth order algorithm* because the function f must be calculated four times in (3.5.14) at each step. Similary, (3.5.8) is a second order algorithm and the Euler algorithm is a first order algorithm. Thus the higher the order, the more calculations must be made at each step, but of course the extra work may improve the accuracy. In fact, for the Runge-Kutta algorithm we find that the truncation error is proportional to h^5. (See [9] page 209 for a proof of this.) However, if we add an extra term to obtain a fifth order algorithm, we find that the truncation error is still proportional to h^5. This is the reason why the Runge-Kutta algorithm is used in the form (3.5.13), and not of an higher order. To show the accuracy of the Runge-Kutta algorithm, compare Table 3.5.1 with Table 3.4.1.

t_i	$y_i = \sqrt{2t_i + 1}$	y_i^* for h_1	y_i^* for h_2	$y_i - y_i^*$ for h_1	$y_i - y_i^*$ for h_2
0.00	1.000000000	1.000000000	1.000000000	0.000000000	0.000000000
0.05	1.048808854	1.048808848	1.048808854	-0.000000005	0.000000000
0.10	1.095445115	1.095445124	1.095445115	-0.000000009	0.000000000
0.15	1.140175425	1.140175436	1.140175425	-0.000000011	0.000000000
0.20	1.183215957	1.183215969	1.183215957	-0.000009012	0.000000000
0.25	1.224744871	1.224744885	1.224744871	-0.000000013	0.000000000
0.30	1.264911064	1.264911078	1.264911064	-0.000000014	0.000000000
0.35	1.303840481	1.303840495	1.303840481	-0.000000014	0.000000000
0.40	1.341640786	1.341640801	1.341640786	-0.000000014	0.000000000
0.45	1.378404875	1.378404889	1.378404875	-0.000000014	0.000000000
0.50	1.414213562	1.414213577	1.414213562	-0.000000014	0.000000000
0.55	1.449137675	1.449137689	1.449137675	-0.000000014	0.000000000
0.60	1.483239697	1.483239711	1.483239697	-0.000000014	0.000000000
0.65	1.516575089	1.516575103	1.516575089	-0.000000014	0.000000000
0.70	1.549193338	1.549193352	1.549193338	-0.000000014	0.000000000
0.75	1.581138830	1.581138844	1.581138830	-0.000000014	0.000000000
0.80	1.612451550	1.612451563	1.612451550	-0.000000013	0.000000000
0.85	1.643167673	1.643167686	1.643167673	-0.000000013	0.000000000
0.90	1.673320053	1.673320066	1.673320053	-0.000000013	0.000000000
0.95	1.702938637	1.702938649	1.702938637	-0.000000013	0.000000000
1.00	1.732050808	1.732050820	1.732050808	-0.000000013	0.000000000

Table 3.5.1: **Fourth order Runge-Kutta algorithm applied to (3.4.12) with $h_1 = 0.05$ and $h_2 = 0.005$**

In practice the Runge-Kutta algorithm is used if only a few y-values to a high accuracy are needed. For example, a few y-values are sometimes needed to start the process in more sophisticated algorithms.

- Read: [32] page 66 or [50] page 393 for more information on the Runge-Kutta algorithm.
- Do: Exercises 17, 18 in §3.7.

3.6 Fruitflies

In §2.2, §2.4, and §2.8 the construction of a suitable population model for a set of experimental data was discussed. Another example of this can be seen in [22], which is a paper by M. E. Gilpin and F. J. Ayala on the population dynamics of two species of fruitflies. They experimented with *Drosophila willistoni* and *Drosophila pseudoobscura* as separate populations and also in competition with each other. Initially they used the logistic model (2.4.1) for the populations, but could not obtain a satisfactory fit on the experimental data. After several models were tried, including ten different models (of which eight were generalizations of the logistic model), they found the following model to be satisfactory (see their paper [4] with J. G. Ehrenfeld):

$$\frac{dN}{dt} = \lambda N \left[1 - \left(\frac{N}{K}\right)^{\theta}\right] \tag{3.6.1}$$

where $N(t)$ denotes the population at time t and λ, K, and θ are constants determined by the experimental data for each species. These constants are shown in Table 3.6.1, where t is measured in days.

Species	K	λ	θ
D. willistoni	1332	1.496	0.35
D. pseudoobscura	791	4.513	0.12

Table 3.6.1: **Model constants for (3.6.1) as determined in [22]**

In the experiments to determine the constants, the initial population α was kept in a 0.24-liter milk bottle with 30 cubic centimeter food at a constant temperature of $21.5°C$. After 1 week the survivors were counted under mild sedation and then transferred to a new bottle. The old bottle was retained because eggs were still hatched in it. After 2 weeks the survivors of both bottles were counted and transferred to a new bottle, whilst both old bottles were retained. This process was continued for five weeks when the original bottle was discarded. From then onwards there were always five bottles in the experiment - as one bottle was discarded weekly, a new bottle took its place. The experiment ran for a year through about fifteen generations of flies.

An interesting conjecture mentioned in the paper on the model (3.6.1) is that the parameter θ always exceeds 1 for vertebrata and is less than 1 for invertebrata. Note that the logistic model (2.4.1) is again obtained when $\theta = 1$.

The differential equation (3.6.1) is nonlinear, and it is not obvious how to determine an explicit solution for an arbitrary value of θ. (We shall discuss this in Project B in §3.8.) Note that the limit value of N as $t \to \infty$ is K. If the initial population α exceeds K, then the initial slope of $N(t)$ is negative and hence N will decrease monotonically to K, as we have seen in §2.5. Similarly, if the initial population α is less than K, then $N(t)$ will increase monotonically to K. An important practical question to ask is how fast the population is increasing; in other words, the time needed to come within a few percent of K. For this purpose the Runge-Kutta algorithm (3.5.13) was used. The model (3.6.1) for the *Drosophila willistoni* is

$$\frac{dN}{dt} = 1.495N\left[1 - \left(\frac{N}{1332}\right)^{0.35}\right], \qquad N(0) = \alpha. \tag{3.6.2}$$

We show the results in Tables 3.6.2 and 3.6.3 and the corresponding graphs for $\alpha = 200$, 500, and 800 in Figure 3.6.1. Note in Table 3.6.3

that the population of flies exceeds 99% of the limit value $K = 1332$ in less than 12 days when $\alpha = 200$, in less than 10 days when $\alpha = 500$, and in less than 8 days when $\alpha = 800$.

t in days	N for $h = 1$	N for $h = 0.5$	N for $h = 0.25$
0	200.00	200.00	200.00
2	588.93	589.06	589.06
4	973.37	973.50	973.51
6	1188.38	1188.50	1188.51
8	1279.02	1279.10	1279.11
10	1313.08	1313.11	1313.11
12	1325.30	1325.33	1325.33
14	1329.64	1329.65	1329.65
16	1331.17	1331.18	1331.18

Table 3.6.2: **The Runge-Kutta algorithm applied to (3.6.2) with $\alpha = 200$**

t in days	N for $\alpha = 200$	N for $\alpha = 500$	N for $\alpha = 800$
0	200	500	800
2	589	908	1102
4	973	1157	1244
6	1188	1267	1300
8	1279	1309	1321
10	1313	1324	1328
12	1325	1329	1331
14	1330	1331	1331

Table 3.6.3: **The Runge-Kutta algorithm applied to (3.6.2) with $h = 1$**

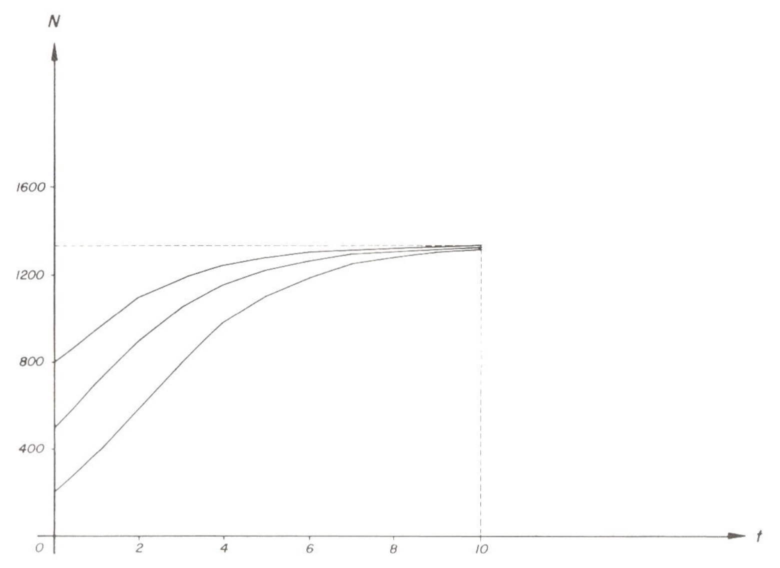

Figure 3.6.1: **Population of fruitflies with different initial populations**

In any numerical approximation the matter of the accuracy of the results should be investigated. In practice the estimate of the error can involve a lot of calculation and in the end produce an estimate which is too crude to be of any practical use. To illustrate this, consider the following estimate (see [56] page 76) for the Runge-Kutta algorithm:

Theorem 3.6.1

Suppose positive constants $A \geq 1$, B, and C exist such that the following estimates for the function f in (3.2.1) hold

$$|f(t,y)| \leq A, \quad \left|\frac{\partial f}{\partial y}\right| \leq B, \quad \left|\frac{\partial^{j+k} f}{\partial t^j \partial y^k}\right| \leq \frac{C}{A^{j+k-1}} \tag{3.6.3}$$

for all non-negative integers j and k such that $j + k \leq 4$. Then the error e_i at $t = t_i$ (excluding round-off errors) for the fourth order Runge-Kutta algorithm (3.5.13) is

$$|e_i| \leq \frac{6AC}{D}(1+C+C^2+C^3+C^4)(e^{D(t_i-t_0)}-1)h^4 \tag{3.6.4}$$

with $D = e^B - 1$ and $0 \leq h \leq 1$.

For the problem (3.6.2) we have

$$f(t,N) = 1.496N\left[1 - \left(\frac{N}{1332}\right)^{0.35}\right]$$

The function f is zero when $N = 0$ and $N = 1332$. Furthermore

$$\frac{\partial f}{\partial N} = 1.496\left[1 - 1.35\left(\frac{N}{1332}\right)^{0.35}\right] \tag{3.6.5}$$

and hence, the slope of f is positive for small values of N, becomes zero at

$$N = 1332(1.35)^{-\frac{1}{0.35}} = 565.10$$

and is negative for larger values of N. Consequently

$$|f(t,N)| \leq 1.496 \times 565.10 \times \left(1 - \frac{1}{1.35}\right) = 219.17 < 220.$$

When $\alpha = 200$ we also have that

$$\begin{aligned}
\left|\frac{\partial f}{\partial N}\right| &\leq 0.524 \qquad \text{(see Exercise 20 in §3.7)} \\
\left|\frac{\partial^2 f}{\partial N^2}\right| &\leq 1.496 \times 1.35 \times 0.35 \times \frac{1}{200} \times (\frac{1332}{1332})^{0.35} = 0.0035 < \frac{0.8}{220} \\
\left|\frac{\partial^3 f}{\partial N^3}\right| &\leq \frac{0.65}{200}\left|\frac{\partial^2 f}{\partial N^2}\right| \leq 0.000011 < \frac{0.6}{(220)^2} \\
\left|\frac{\partial^4 f}{\partial N^4}\right| &\leq \frac{1.65}{200}\left|\frac{\partial^3 f}{\partial N^3}\right| \leq 0.00000009 < \frac{1.01}{(220)^3}
\end{aligned}$$

Since the derivatives of f with respect to t are all zero, the best choices, for the constants in (3.6.3) are in this case

$$A = 220, \; B = 0.524, \text{ and } C = 1.01.$$

and then $D = e^{0.524} - 1 < 0.689$. Hence we have in (3.6.4) at $t = 10$ the estimate

$$|e_i| \leq \frac{6 \times 220 \times 1.01}{0.689} \times 5.101 \times (e^{6.89} - 1)h^4 < 9686750h^4 \qquad (3.6.6)$$

According to this estimate, we must take $h = 0.01$ to ensure accuracy to the nearest fly. (Note that the estimate on the constant D is crucial since it appears in the exponent in the brackets.) If this value of h is compared to Table 3.6.2 we immediately see that the estimate is far too crude.

When the formula for an estimate is too involved, it is a common practice to repeat the algorithm for three different step lengths, usually h, $\frac{1}{2}h$, and $\frac{1}{4}h$, and then compare the values of the solution. If the answers agree in all three cases to the required number of decimal places, it is usually accepted as correct to this number of decimal places. This was done in Table 3.6.2, and by this convention it is sufficient to take $h = 1$ for results accurate to the nearest fly. (From a practical viewpoint we are not really interested in fractions of a fly!) Hence $h = 1$ was used in Table 3.6.3.

The rationale behind this practice is the following: Suppose the error of the algorithm can be expressed as

$$e = Kh^p \qquad (p > 0) \qquad (3.6.7)$$

and let y_l, y_2 and e_l, e_2 denote the approximate values and errors obtained with h and $\frac{1}{2}h$, respectively. Let y denote the exact solution of the boundary value problem.

$$\begin{aligned} y_1 - y_2 &= (y + e_1) - (y + e_2) \\ &= e_1 - e_2 \\ &= Kh^p - K\left(\frac{h}{2}\right)^p \\ &= \left(1 - \frac{1}{2^p}\right) Kh^p \\ &= \left(1 - \frac{1}{2^p}\right) (\text{error due to step length } h) \end{aligned}$$

Hence, the difference between the approximate values is an indication of the error. Note, however, that this argument is based on the assumption (3.6.7). In practice, however, we usually have only an inequality (with $K > 0$) available; and the above deduction is then not valid.

One final word of warning: never accept answers from a numerical algorithm in good faith. Be certain that a solution exists and check your answers as far as possible. Blind acceptance without common sense is a game of Russian roulette - one day it won't be a click but a big bang and your reputation as a mathematician will be gone forever.

- Do: Exercises 19, 20 in §3.7.
- Do: Project B in §3.8.

3.7 Exercises

(1) Solve the boundary value problem

$$\begin{cases} \dfrac{dy}{dt} = \sqrt{y} & (t > 0) \\ y(0) = 0 \end{cases}$$

Show that there are two solutions to this problem. Explain how this is possible in the light of Theorem 3.2.1 (or Theorem 2.4.1).

(2) Solve the boundary value problem

$$\begin{cases} \dfrac{dy}{dt} = ty^2 & (t > 0) \\ y(0) = 1 \end{cases}$$

Determine the interval on the t-axis where the solution exists.

(3) Apply Theorem 3.2.1 to Exercise 2 and determine the largest value of δ so obtained. What is the bound on y when $|t| \leq \delta$? Draw the graph of the solution y as found in Exercise 2, and show on it the rectangle R of Theorem 3.2.1 for this largest δ.

(4) Apply the Euler algorithm to Exercise 2 with $h = 0.1$ on the interval $[0, 0.4]$. Show both the values of the solution and the approximation by the Euler algorithm at the five points in the interval in a table.

(5) Repeat Exercise 4 with $h = 0.05$ (i.e., with nine points in the interval $[0, 0.4]$) to see if a smaller step length improves the approximation. Draw the graph of the solution y and the two approximations on the same axes.

(6) Solve the initial value problem

$$\frac{dy}{dt} + y = t + 1, \qquad y(0) = 1$$

(7) Run the program EULER for the initial value problem in Exercise 6 with $h = 0.1$ and $n = 20$. Compare the approximate values with the exact solution found in Exercise 6.

(8) Repeat Exercise 4 with $h = 0.02$.

(9) The round-off error is dependent on the sequence in which arithmetical operations are executed. For example, the distributive law $a(b - c) = ab - ac$ does not hold when the multiplications are rounded off. Show this with $a = 0.111, b = 2.060$, and $c = 1.030$ by rounding off after each multiplication to three decimal places.

(10) If b and c are almost equal and products are rounded off, which is better to minimize the round-off errors: to calculate $a(b - c)$ or $ab - ac$? Illustrate your answer with a numerical example.

(11) Use the Euler algorithm with $h = 0.05$ to calculate $y(1)$ where y is the solution of

$$\begin{cases} \dfrac{dy}{dt} = y - \dfrac{2t}{y} & (t > 0) \\ y(0) = 1. \end{cases}$$

Repeat Table 3.4.1 for this problem. (Note that the solution to this boundary value problem is also the solution of (3.4.12).) Show that in this case (3.4.11) leads to the estimate $|e_i| \leq 28h$.

(Hint: Since $y^{(1)}(0) = 1$ and $y(0) = 1$, assume that $1 \leq y \leq 2$ in the closed interval $[0, 1]$ to do the necessary estimates, and check afterwards in the table that y is, in fact, less than 2.)

(12) Use the Euler algorithm with $h = 0.1$ to calculate $y(1)$ where y is the solution of

$$\begin{cases} \dfrac{dy}{dt} = y^2 & (t > 0) \\ y(0) = 1. \end{cases}$$

Solve the boundary value problem. Repeat Table 3.4.1 for this problem. How could you have anticipated this crisis if you had not known the solution?

(13) In a study of the rare fruitfly *Drosophila gansobscura* it was decided to model the population N at time t as follows:

$$\begin{cases} \dfrac{dN}{dt} = 0.2N\left(1 - N^{0.45}\right) & (0 < t < \infty) \\ N(0) = 5 \end{cases}$$

where t is measured in days and the population in hundreds.

(a) Use the Euler algorithm to calculate the approximate population each day for 3 days. (Use h = 1 and calculate to the nearest fly.)

(b) Show that the solution $N(t)$ is always more than 1 and less than 5. (You may assume that the solution is unique. Hint: look at §2.5 if you have forgotten.)

(c) Estimate the error by (3.4.11) after 3 days. How small must we choose h to ensure that $N(3)$ is correct to the nearest fly? (Ignore the round-off error.)

(14) You are the applied mathematician on a team of scientists who are investigating the distribution of seeds of a plant family in the Richtersveld, a very arid area in Africa. Since the seeds are carried by the wind, it is important to determine the speed v of the seeds at any time instant t as the seeds drops vertically. You decide to measure this in the laboratory with the aid of a stroboscope and find that

t in seconds	v in meters per second
0	0.48
0.1	1.16
0.2	1.60
0.3	1.98
0.4	2.36

(a) Construct a model which describes v as a function of t by assuming that the air resistance is directly proportional to some power of v. (Look at §2.9 again if you had forgotten!) If $g = 9.81\ ms^{-2}$, the initial acceleration $6.8\ ms^{-2}$, and the terminal speed $9\ ms^{-1}$ are known, show that

$$\frac{dv}{dt} = 9.81 - 4.07v^{0.4}.$$

(b) Use the Euler algorithm to calculate v from the differential equation in (a), and compare your answers with the above table. (Assume that the solution exists on the closed interval $[0, 0.4]$.)

(c) Estimate the error at $t = 0.4$ with the aid of the formula (3.4.11). (Ignore the round-off error.)

(15) Repeat the table of Exercise 12 when the modified Euler algorithm is used.

(16) Repeat the table of Exercise 12 when the Heun algorithm is used.

(17) Repeat the table of Exercise 12 when the Runge-Kutta algorithm is used.

(18) Write a program to implement the Runge-Kutta algorithm. Use your program to obtain a table of y-values for Exercise 11. Compare this result with the result of Exercise 11.

(19) Do the estimate (3.6.4) at $t_i = 10$ for the case when $\alpha = 800$ in (3.6.2) and compare it with (3.6.6).

(20) Good estimates of A, B, and C in the application of (3.6.4) are very important.

(a) Use absolute values in (3.6.5) and utilize the fact that $N < 1332$ to show that we can take $B = 3.6$. Calculate also then D, and investigate the effect of the exponential term in (3.6.4).

(b) To find a better estimate in (3.6.5), draw a rough sketch of the function in square brackets. Calculate the values at $N = 200$ and $N = 1332$. Show that the estimate

$$\left| \frac{\partial f}{\partial N} \right| \leq 0.524$$

used in §3.6 is indeed valid.

3.8 Projects

Project A^*: The Motion of an Experimental Rocket

The company **BAM** (which is an acronym for "Ballistic Armament Models") is developing a short-range one stage rocket, called the *Lance Cloudprobe*, to be used in weather research. The rocket will be launched vertically. It is very important for research purposes to know how high the rocket will go.

Let the velocity of the rocket be denoted by v and the time by t and measure these quantities in meters per second and seconds, respectively, where $t = 0$ is chosen at ignition. During the time interval of 0.5 seconds the fuel is burnt up, and $v(t)$ is given by

$$\frac{dv}{dt} = -9.8 + \frac{302.3 - 72.5 \times 10^{-6}\rho v^2}{1 - 0.7692t} \qquad (0 \leq t \leq 0.5) \qquad (3.8.1)$$

where ρ denotes the density of air in SI units.

(1) The motion of a rocket was discussed in §2.11 with certain simplifying assumptions. Compare (3.8.1) with (2.11.8) and discuss the similarities and differences between the two models.

(2) Assume that the rocket is launched in a vacuum to simplify (3.8.1). Integrate the simplified equation to obtain the velocity of the rocket at $t = 0.5$ when the fuel is exhausted.

(3) Write a computer program to calculate $v(0.5)$ from (3.8.1) by the Euler algorithm. Use $\rho = 0$ to compare $v(0.5)$ with the result in (2). Use $\rho = 1.225 kg/m^3$ for the case of air resistance. Do three calculations with step lengths of (i) 0.05, (ii) 0.01, (iii) 0.005 in each case.

(4) For $t > 0.5$ the rocket is a free falling body, as discussed in §2.9. Show that (3.8.1) then becomes

$$\frac{dv}{dt} = -9.8 - 117.8 \times 10^{-6}\rho v^2 \qquad (0.5 \leq t \leq T) \qquad (3.8.2)$$

until the time instant T when $v = 0$. Calculate T.

(5) Let $y(t)$ denote the altitude of the rocket at time t. Expand your computer program in (3) to include the calculation of the altitude at each step of the Euler algorithm. Obtain then approximately the maximum altitude of the *Lance Cloudprobe* rocket.

Project B: An Analytic Look at Fruitflies

In the paper [23] S. P. Gordon shows that a differential equation of the type

$$\frac{dy}{dt} = a\left(y - by^{p+1}\right) \tag{3.8.3}$$

can be solved analytically by writing

$$\frac{1}{y(1 - by^p)} = \frac{1}{y} + \frac{by^{p-1}}{1 - by^p}.$$

(1) Solve (3.8.3) by this method, using $y(0) = \alpha$ as the initial value.

(2) Use (1) to find the solution of (3.6.2).

(3) Calculate the values of $N(t)$ at $t = 0, 2, 4, 6, 8, 10, 12, 14, 16$ and compare your results with Table 3.6.2.

(4) Differentiate (3.8.3) with respect to t to obtain the y-value of the point of inflection. Then use the solution in (1) to find the corresponding t-value of the point of inflection.

(5) Suppose a table of data is given and you can see from this data that the point of inflection is at (T, B) and the limit value as $t \to \infty$ is A. Show that p is given by

$$p + 1 = \left(\frac{B}{A}\right)^p$$

and once p is known, then

$$b = A^{-p} \qquad \text{and} \qquad a = \frac{1}{pT}\ln\left(\frac{1 - b\alpha^p}{bp\alpha^p}\right).$$

(6) Consider the population of the town Huntington as given in [23]:

t	N
1900	9,483
1910	12,004
1920	12,893
1930	25,582
1940	31,768
1950	47,506
1960	126,221
1970	199,486
1980	201,512

Plot the population by hand on graph paper, and try to estimate the point of inflection and the limit value of the population as $t \to \infty$. Then use (5) to obtain a model of the type (3.8.3) which will approximate this data.

3.9 Mathematical Background

This section contains the proof of the very important Theorem 3.2.1. Note especially where the conditions of the theorem are used in the proof. It may happen that you have to do a problem which does not satisfy these conditions, but which is close to it; and then you will have to devise a new theorem and proof for your problem by adapting the statement and proof of Theorem 3.2.1 to suit your situation. To refresh your memory we also look very briefly at those numerical procedures for the approximation of definite integrals referred to in §3.3.

Theorem 3.2.1

Let $f(t,y)$ be continuous on the rectangle $R = \{(t,y) : |t-t_0| \leq a, |y-y_0| \leq b\}$ and let $f(t,y)$ satisfy a Lipschitz condition in y on the closed interval $[y_0 - b, y_0 + b]$ for each t in the closed interval $[t_0 - a, t_0 + a]$. Then there exists a unique solution of the boundary value problem

$$\frac{dy}{dt} = f(t,y), \qquad y(t_0) = y_0 \tag{3.9.1}$$

on the interval $I_\delta = \{t : |t - t_0| \leq \delta\}$ where $\delta =$ minimum $\{a, \frac{b}{M}\}$ with M the maximum of $|f(t,y)|$ on R.

Proof of Theorem 3.2.1

The proof follows the idea of E. Picard to approximate the solution by successive integrals. Note that if $y(t)$ satisfies

$$y(t) = y_0 + \int_{t_0}^{t} f(s, y(s))\, ds \tag{3.9.2}$$

then $y(t)$ is also a solution of (3.9.1). Define the operator

$$F[y](t) = y_0 + \int_{t_0}^{t} f(s, y(s))\, ds \tag{3.9.3}$$

which maps a function of t onto another function of t. If $x(t)$ is any function continuous on the interval I_δ, with $x(t_0) = y_0$ and $|x(t) - y_0| \leq b$

on I_δ, then $F[x](t)$ will have the same three properties. The continuity and value at $t = t_0$ follow immediately from (3.9.3). The third property is a consequence of the bound M on the function $|f|$ on the domain R:

$$\begin{aligned}
|F[x](t) - y_0| &= \left|\int_{t_0}^{t} f(s, x(s))\, ds\right| \\
&\le \int_{t_0}^{t} |f(s, x(s))|\, ds \qquad (t \ge t_0) \\
&\le \int_{t_0}^{t} M\, ds \\
&\le M|t - t_0| \\
&\le b \qquad (t \in I_\delta)
\end{aligned}$$

Now define successively

$$\begin{aligned}
x_0(t) &= y_0 \\
x_1(t) &= F[x_0] \\
x_2(t) &= F[x_1] \\
&\vdots \\
x_n(t) &= F[x_{n-1}].
\end{aligned}$$

For any integer n, the functions $x_n(t)$ are continuous on the interval I_δ, with $x_n(t_0) = y_0$ and $|x_n(t) - y_0| \le b$ on I_δ. We shall show that the sequence $\{x_n(t)\}_{n=0}^{\infty}$ converges uniformly on the interval I_δ to the required solution $y(t)$ of (3.9.2). (See [16] page 529 for a discussion of uniform convergence.)

By the Lipschitz condition there exists a constant k such that

$$|f(t, y_1) - f(t, y_2)| < k|y_1 - y_2|$$

for all $t \in [t_0 - a, t_0 + a]$ and y_1 and y_2 in the interval $[y_0 - b, y_0 + b]$. Hence

$$\begin{aligned}
|x_n(t) - x_{n-1}(t)| &= |F[x_{n-1}](t) - F[x_{n-2}](t)| \\
&= \left|\int_{t_0}^{t} \{f(s, x_{n-1}(s)) - f(s, x_{n-2}(s))\}\, ds\right| \\
&\le \int_{t_0}^{t} |f(s, x_{n-1}(s)) - f(s, x_{n-2}(s))|\, ds \\
&< k\int_{t_0}^{t} |x_{n-1}(s) - x_{n-2}(s)|\, ds \qquad (3.9.4)
\end{aligned}$$

We now show by induction on n that

$$|x_n(t) - x_{n-1}(t)| < \frac{Mk^{n-1}}{n!}(t - t_0)^n \quad \text{for } n = 1, 2, 3, \ldots \tag{3.9.5}$$

For n = 1 we have

$$\begin{aligned} |x_1(t) - x_0(t)| &= \left| \int_{t_0}^{t} f(s, y_0)\, ds \right| \\ &\leq \int_{t_0}^{t} |f(s, y_0)|\, ds \\ &\leq \int_{t_0}^{t} M\, ds \\ &= M(t - t_0) \end{aligned}$$

Suppose (3.9.5) holds. Then by (3.9.4)

$$\begin{aligned} |x_{n+1}(t) - x_n(t)| &< k \int_{t_0}^{t} |x_n(s) - x_{n-1}(s)|\, ds \\ &< k \int_{t_0}^{t} \frac{Mk^{n-1}}{n!}(s - t_0)^n\, ds \\ &= \frac{Mk^n}{(n+1)!}(t - t_0)^{n+1} \end{aligned}$$

which shows that (3.9.5) holds for all positive integers n.

Note that

$$x_n(t) = y_0 + \sum_{i=1}^{n} \{x_i(t) - x_{i-1}(t)\}$$

Since the infinite series

$$M \sum_{n=1}^{\infty} \frac{k^{n-1}}{n!}(t - t_0)^n = \frac{M}{k}\left(e^{k(t-t_0)} - 1\right)$$

converges uniformly for $t \in [t_0 - a, t_0 + a]$, it follows by the comparison test (see [64] page 285 or [16] page 520) and (3.9.5) that the infinite series

$$\sum_{n=1}^{\infty} \{x_n(t) - x_{n-1}(t)\}$$

will also converge uniformly for $t \in I_\delta$. Hence, the sequence $\{x_n(t)\}_{n=1}^{\infty}$ is uniformly convergent on I_δ.

Denote the limit of this sequence by $x(t)$. Since

$$x_n(t) = F[x_{n-1}]$$

it follows for any $t \in I_\delta$ that

$$\begin{aligned}\left| x(t) - y_0 - \int_{t_0}^{t} f(s, x(s))\, ds \right| &\leq |x(t) - x_n(t)| \\ &\quad + \int_{t_0}^{t} |f(s, x(s)) - f(s, x_{n-1}(s))|\, ds \\ &\leq |x(t) - x_n(t)| \\ &\quad + k \int_{t_0}^{t_0+\delta} |x(s) - x_{n-1}(s)|\, ds\end{aligned}$$

By the uniform convergence of $\{x_n(t)\}_{n=1}^{\infty}$ to $x(t)$, the right hand side can be made arbitrarily small by choosing n large enough. This shows that $x(t)$ is, in fact, a solution of (3.9.2) and hence a solution of (3.9.1).

Numerical Integration

The integral

$$I = \int_a^b f(x)\, dx$$

can be seen as the area of a planar configuration of which three sides are straight lines and the rest of the boundary is the graph of the function f (provided that the concept of negative area is introduced). If $n + 1$ equidistant points t_i with $t_0 = a$ and $t_n = b$ are chosen on the x-axis and we write $h = t_{i+1} - t_i$, then the following algorithms to calculate I can easily be interpreted geometrically:

(1) *The left endpoint rectangle rule*:

Draw n rectangles of which the first one has height $f(a)$, the second one height $f(t_1)$, etc., and all the rectangles have a common width h. Then the sum of the area of the rectangles is approximately the area under the graph of the function f:

$$I \approx h[f(a) + f(t_1) + \ldots + f(t_{n-1})]$$

(2) *The trapezium rule*:

Draw n trapeziums by connecting the successive points $(t_i, f(t_i))$ and $(t_{i+1}, f(t_{i+1}))$ by straight line segments. Again the sum of the areas of all the trapeziums is approximately the area under the graph f:

$$I \approx \frac{h}{2}[f(a) + 2f(t_1) + \ldots + 2f(t_{n-1}) + f(b)]$$

(3) *Simpson's rule*:

For this rule n must be an even integer. Draw a parabola through the first three successive points $(a, f(a))$, $(t_1, f(t_1))$, and $(t_2, f(t_2))$. Draw another parabola through the next three successive points, and continue to do this until the graph of f is approximated on the interval $[a, b]$ by $\frac{n}{2}$ parabolas. When the sum of the area under the parabolas is calculated (see [16] page 483), the following rule results:

$$I \approx \frac{h}{6}[f(a)+4f(t_1)+2f(t_2)+4f(t_3)+2f(t_4)+\ldots+4f(t_{n-1})+f(b)]$$

4

Laplace Transforms

4.1 Linear Transforms

The Laplace transform is a powerful method for the solution of linear differential equations. In this chapter we shall not be interested in covering the vast theory of Laplace transforms, but rather in looking at the Laplace transform as a very handy tool to solve boundary value problems. To this end we define the Laplace transform, derive some of its properties, and apply it to the solution of ordinary linear differential equations. In the subsequent chapters this method of solution will be used many times in the mathematical models presented there.

Let us first look at the concept of a *linear transform*. A function like $y = 2x + 3$ maps the set of real numbers onto the set of real numbers, which can be shown graphically with the aid of two orthogonal axes. For example, the number 2 (on the x-axis) is mapped on the number 7 (on the y-axis). An obvious extension of the "mapping" idea is to map a class of functions onto another class of functions, instead of a set of numbers onto another set of numbers. Think, for example, of the well-known process of differentiation. The function

$$f = \{(t, y) : \; y = t^2\}$$

is mapped on the function

$$Df = \{(t, y) : \; y = 2t\}.$$

This operator D has the important property that it is a *linear transformation*: that means that D maps a certain class of functions onto another class of functions in such a way that

$$\left.\begin{array}{rcl} D(f+g) & = & Df + Dg \\ D(\alpha f) & = & \alpha Df \end{array}\right\} \qquad (4.1.1)$$

where α is any real number. (Linear operators can similarly be defined on complex functions, in which case α would be a complex number.)

There are many more examples of linear transformations which map one class of functions onto another class of functions. One popular way to define linear transformations is to use integrals in the following way:

$$T[f](s) = \int_a^b K(s,t)f(t)\,dt \tag{4.1.2}$$

The function $K(s,t)$ is known as the *kernel* of the transformation T. The function $T[f]$ (which is a function of s) is called the *transform* of f (which is a function of t). By the properties of integrals it follows immediately that T is always a linear transformation (see Exercise 1 in §4.5). For a given kernel $K(s,t)$, the set of functions f for which the integral in (4.1.2) exists must be determined to obtain the domain of the operator T.

The importance of integral transforms is that, in certain circumstances, they enable us to solve a problem after it has been transformed. Let us illustrate this with a well-known analogy where sets of real numbers are mapped onto sets of real numbers. Suppose the problem

$$x^5 = 2.364$$

must be solved for the variable x. By the log-function we transform this problem to a new (transformed) problem in the variable $logx$, which can easily be solved. Then we use the inverse function to transform back to the original variable x. This procedure is illustrated in Figure 4.1.1. Note that it is the special property $logx^n = nlogx$ of the logarithmic function which enabled us to solve the transformed problem.

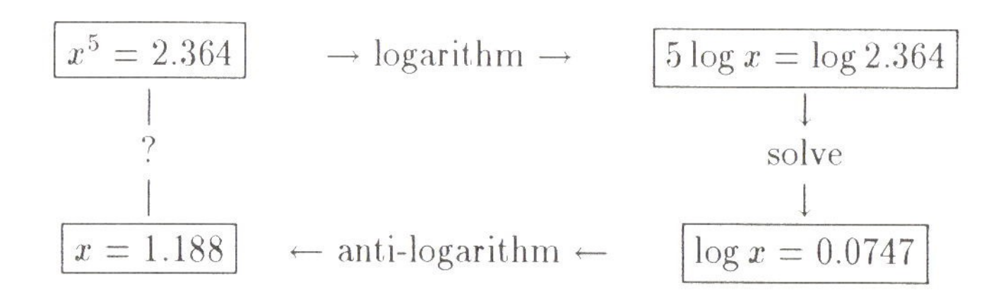

Figure 4.1.1: **Illustration of a transformed problem in real numbers**

In the same way integral transformations are used to solve boundary value problems. A boundary value problem defined on the t-domain (i.e., with t as the independent variable) is transformed to the s-domain where it is solved, and the solution is then found with the aid of the inverse transformation.

There are many of these integral transformations, each with its own special properties. For example, when $a = 0$, $b = \infty$, and $K(s,t) = t^s$, we call (4.1.2) the Mellin transform; or when $a = 0$, $b = \pi$, and $K(s,t) = \sin st$, we call (4.1.2) the modified finite Fourier sine transform; etc. In this chapter we shall consider a very popular transformation, namely the Laplace transform when $a = 0$, $b = \infty$, and $K(s,t) = e^{-st}$. The main reason for our interest in this transformation will become clear when we come to Theorem 4.2.2.

Definition 4.1

The Laplace transform of a function $f(t)$ is the function

$$L[f] = F(s) = \int_0^{\infty} e^{-st} f(t)\, dt \tag{4.1.3}$$

The variable s is in general a complex variable and $F(s)$ a complex function, but we shall consider s only as a real variable. The reason for this is that we shall use the Laplace transform only in the special case of a solution method for ordinary linear differential equations. It must be understood that we shall only look at the tip of the iceberg of the theory and applications of the Laplace transform (see, for instance, a book like [15]).

Example

Let $f = \{(t, y)\colon y = 1,\ t \geq 0\}$. Then by the definition of the Laplace transform

$$\begin{aligned} L[f] &= \int_0^{\infty} e^{-st}\, dt \\ &= \lim_{T \to \infty} \int_0^{T} e^{-st}\, dt \\ &= \lim_{T \to \infty} \left(-\frac{1}{s} e^{-sT} + \frac{1}{s} \right) \qquad (s \neq 0) \\ &= \frac{1}{s} \qquad (s > 0) \end{aligned}$$

Hence the function f in the t–domain is mapped onto the function $F = \{(s, y)\colon y = \frac{1}{s},\ s > 0\}$ in the s-domain.

Note also that f need only be defined on the interval $[0, \infty)$. *We shall always consider f to be zero on the negative t-axis, even if it was originally defined otherwise.* This is necessary in the cases where f is translated to the right. In particular it is needed to obtain the following nice property of Laplace transforms:

$$L[f(t-a)] = e^{-sa} L[f] \tag{4.1.4}$$

This property is important when Laplace transforms are applied to time-delay equations. (We have seen an example of a time-delay equation in the potato blight model of §2.8.) It is a simple matter to show why (4.1.4) holds:

$$\begin{aligned} L[f(t-a)] &= \int_0^\infty e^{-st} f(t-a)\, dt && \text{by (4.1.3)} \\ &= \int_{-a}^\infty e^{-s(\tau+a)} f(\tau)\, d\tau && \text{by } \tau = t-a \\ &= e^{-sa} \int_0^\infty e^{-s\tau} f(\tau)\, d\tau && \text{by } f(\tau) = 0 \text{ on } [-a, 0) \\ &= e^{-sa} L[f] && \text{by (4.1.3)} \end{aligned}$$

Note the essential part played in the third line by the convention that f is zero on the negative t-axis.

- Do: Exercises 2, 3, 4, 5 in §4.5.

4.2 Some Properties of Laplace Transforms

You will see in §4.3 that Laplace transforms are very handy in solving boundary value problems. To use this tool efficiently you must know its properties and its limitations. We have already shown in §4.1 that the Laplace transform is linear:

$$L[\alpha f + \beta g] = \alpha L[f] + \beta L[g] \tag{4.2.1}$$

where α and β are real numbers and f and g are functions for which the Laplace transforms exist. On the other hand we do not have a similar property for the product of two functions.

First of all, the class of functions which are transformable and which are mapped in a one-to-one manner by the Laplace transform, must be determined - the latter property is especially important since we need the inverse transform to return to the original problem in the t-domain (see Figure 4.1.1). To define this class of functions, we need four important definitions. It is essential that these definitions are well understood.

Definition 4.2

The **left limit** $f(a^-)$ *and* **right limit** $f(a^+)$ *of a function* f *at the point* $t = a$ *are, respectively*

$$f(a^-) = \lim_{\substack{t \to a \\ t<a}} f(t) \quad \text{and} \quad f(a^+) = \lim_{\substack{t \to a \\ t>a}} f(t)$$

Definition 4.3

A function f *is* **piecewise continuous** *on the closed interval* $[a, b]$ *if* f *is continuous on* $[a, b]$ *with the exception of a finite number of points* $\{t_i\}_{i=1}^n$ *in the interval, where the left and right limits* $f(t_i^-)$ *and* $f(t_i^+)$ *must exist.*

Definition 4.4

A function f *is of* **exponential order** α *if there exist positive real numbers* T, c, *and* α *such that*

$$|f(t)| < ce^{\alpha t} \qquad (t > T)$$

Definition 4.5

A function f *is of* **class E** *if (i) the function is defined on the interval* $0 \leq t < \infty$, *(ii) is of exponential order, (iii) is piecewise continuous on every finite interval, and (iv) is defined* $f(t_i) = \frac{1}{2}[f(t_i^+) + f(t_i^-)]$ *at each point of discontinuity.*

Examples

(1) The function $f = \{(t, y)\colon\ y = \tan t,\ 0 \leq t \leq \pi\}$ is discontinuous, and the left and right limits do not exist at $t = \frac{\pi}{2}$. Hence, f is not piecewise continuous.

(2) The function f defined by

$$f(t) = \begin{cases} 0 & (0 \leq t \leq 1) \\ 3 & (1 < t \leq 2) \end{cases}$$

is piecewise continuous on $[0, 2]$. There is only one discontinuity at $t = 1$ where $f(1^-) = 0$ and $f(1^+) = 3$. Note that f is not of class E because $f(1) = 0$.

(3) The function f defined by

$$f(t) = \begin{cases} 0 & (0 \le t < 1) \\ 1.5 & (t = 1) \\ 3 & (t > 1) \end{cases}$$

is of class E.

(4) The function $f = \{(t, y): y = t^n,\ n \text{ any integer},\ t \ge 0\}$ is of exponential order. If $n < 0$, the statement is obviously valid because the function is bounded. If $n \ge 0$, let α be an arbitrarily small positive number. Since for any $t \ge 0$ (see §2.14)

$$e^{\alpha t} = 1 + \alpha t + \ldots + \frac{\alpha^n t^n}{n!} + \ldots$$

it follows immediately that

$$t^n < n!\alpha^{-n} e^{\alpha t} \qquad (t \ge 0)$$

Since f is continuous, it then also follows that f is of class E.

(5) The function $f = \{(t, y): y = e^{t^2},\ t \ge 0\}$ is not of exponential order. Take arbitrarily large numbers c and α. Choose for T any number exceeding both $\sqrt{2 \ln c}$ and 2α. Then for $t > T$

$$e^{t^2} = e^{\frac{1}{2}t^2}.e^{\frac{1}{2}t^2} > ce^{\alpha t}$$

- Do: Exercises 6, 7, 8, 9, 10, 11, 12 in §4.5. Note in Exercise 11 that if f and g are of class E, then any linear combination of f and g will also be of class E.

We can now state a theorem which describes the class of functions which can be transformed in a one-to-one manner.

Theorem 4.2.1

If f is of class E with exponential order α, then the Laplace transform of f exists for all $s > \alpha$. Moreover, $\lim_{s \to \infty} L[f] = 0$. Two functions of class E which are not equal have different Laplace transforms.

Note that the functions of class E are not the only functions which are Laplace transformable (see, for example, Exercises 5 and 10(e) in §4.5).

However, the set of functions of class E is sufficient for our purposes in this book.

Theorem 4.2.2

If f is continuous on the interval $[0, \infty)$ and of exponential order α, and the derivative f' of f is continuous on the interval $(0, \infty)$, then the Laplace transform of f' exists for $s > \alpha$ and

$$L[f'] = sL[f] - f(0) \tag{4.2.2}$$

Theorem 4.2.2 is also valid when f' is piecewise continuous (see Exercise 13 in §4.5). Higher order derivatives can be transformed in the same way:

$$\begin{aligned} L[f''] &= sL[f'] - f'(0) \\ &= s^2 L[f] - sf(0) - f'(0) \end{aligned}$$

and in general, for any positive integer n, if the nth derivative $f^{(n)}$ is continuous, and all the lower order derivatives are of exponential order α and continuous on $[0, \infty)$, then for $s > \alpha$

$$L[f^{(n)}] = s^n L[f] - \sum_{i=1}^{n} s^{n-i} f^{(i-1)}(0) \tag{4.2.3}$$

This property of the Laplace transform is the real reason for the inclusion of this chapter in the book, because it implies that an ordinary linear differential equation with constant coefficients will be transformed into an equation without derivatives (as we shall see in §4.3). Before we do some examples to illustrate this, let us briefly look at some other properties of the Laplace transform.

The first property shows that the integral of the function f is transformed by dividing the transform of f by s in contrast to (4.2.2) where the transform of f is multiplied by s. The other two properties are both transforms of products of special functions.

Theorem 4.2.3

If the function f is of class E with exponential order α, then so is the integral of f and for $s > \alpha$

$$L\left[\int_0^t f(\sigma)\, d\sigma\right] = \frac{1}{s} L[f] \qquad (4.2.4)$$

Moreover, all the derivatives of $L[f]$ exist and for any positive integer n and $s > \alpha$

$$L[t^n f(t)] = (-1)^n \frac{d^n}{ds^n} L[f] \qquad (4.2.5)$$

For any real number a, $s > \alpha + a$, and $F(s) = L[f]$,

$$L[e^{at} f(t)] = F(s - a) \qquad (4.2.6)$$

- Do: Exercises 13, 14, 15 in §4.5.

In contrast to (4.2.1) where the transform of the sum of two functions was shown to be the sum of the transforms of the functions, there is no similar formula for the product of two functions. The best we can do is to introduce a new type of product.

Definition 4.6

The **convolution** *of two functions f and g is the integral*

$$f * g(t) = \int_0^t f(\tau) g(t - \tau)\, d\tau \qquad (4.2.7)$$

provided that the integral exists for $t > 0$.

We shall show in the next theorem that this product is commutative, associative, and distributive, just like ordinary multiplication. It also has the important property that the Laplace transform of a convolution of two functions is equal to the product of the Laplace transforms of these two functions.

Theorem 4.2.4

If f, g, and h are any piecewise continuous functions on a closed interval $[0, T]$, then $f * g(t)$ is continuous on $[0, T]$, and

(*i*) $f * g = g * f$
(*ii*) $f * (\alpha g) = \alpha(f * g)$ for any real number α
(*iii*) $(f * g) * h = f * (g * h)$
(*iv*) $f * (g + h) = f * g + f * h$

If f and g are of class E with exponential order α, then for $s > \alpha$

$$L[f * g] = L[f]L[g] \tag{4.2.8}$$

This theorem is especially important in mathematical models where a forcing term appears (in other words a term which does not include the dependent variable). In these models the transformed equation simplifies to a product of two Laplace transforms on the right hand side, so that (4.2.8) must be applied before the inverse transform can be taken.

- Do: Exercises 16, 17 in §4.5.

4.3 Solving Differential Equations with Laplace Transforms

Laplace transforms are included in this book because they supply an easy, efficient, and quick procedure to find the solution of a linear boundary value problem: transform the boundary value problem in the t-domain to the s-domain, solve the linear equation in the s-domain, and use the inverse transform to obtain the answer in the t-domain. To illustrate this procedure, a few examples will be shown in this section.

Example 1

$$\frac{dy}{dt} = 4y + 9te^t \qquad y(0) = 1 \tag{4.3.1}$$

Assume that the solution is of class E. Then the transform $Y(s) = L[y]$ exists by Theorem 4.2.1. By the definition in §1.2 the solution $y(t)$ is continuous, and hence by (4.3.1) the derivative $\frac{dy}{dt}$ is also continuous for $t > 0$. Now Theorem 4.2.2 and (4.2.1) can be used to transform (4.3.1) to

$$\begin{aligned} sY(s) - 1 &= 4Y(s) + 9L[te^t] \\ &= 4Y(s) + 9(s-1)^{-2} \end{aligned}$$

where Table 4.4.1, number 3, was used. Solve for Y in the s-domain

$$Y(s) = \frac{1}{s-4} + \frac{9}{(s-1)^2(s-4)}$$

and use the inverse transform to return to the t-domain:

$$y(t) = L^{-1}\left[\frac{1}{s-4}\right] + L^{-1}\left[\frac{9}{(s-1)^2(s-4)}\right] \tag{4.3.2}$$

By Table 4.4.1, number 2, the inverse transform of first term is:

$$L^{-1}\left[\frac{1}{s-4}\right] = e^{4t}$$

The expression in the square brackets of the second term in (4.3.2) is a product of two transforms. Hence the inverse transform can be found with the aid of a convolution as in (4.2.8) (see Exercise 17(b) in §4.5). An alternative method is to use partial fractions

$$\frac{9}{(s-1)^2(s-4)} = \frac{1}{s-4} - \frac{1}{s-1} - \frac{3}{(s-1)^2}$$

Since L is linear, L^{-1} must also be linear, and so

$$\begin{aligned} L^{-1}\left[\frac{9}{(s-1)^2(s-4)}\right] &= L^{-1}\left[\frac{1}{s-4}\right] - L^{-1}\left[\frac{1}{s-1}\right] \\ &\quad -3L^{-1}\left[\frac{1}{(s-1)^2}\right] \\ &= e^{4t} - e^t - 3te^t \end{aligned}$$

where Table 4.4.1 was used again. By (4.3.2) we thus have the solution of (4.3.1):

$$y(t) = 2e^{4t} - e^t - 3te^t$$

This solution is, in fact, of class E (remember the assumption just after (4.3.1) at the start of the problem!), and so a solution of (4.3.1) exists. Moreover, by Theorem 2.4.1 the solution is unique, which means that the above solution is the only solution.

Example 2

$$\left.\begin{array}{ll}\dfrac{d^2y}{dt^2} - 2\dfrac{dy}{dt} + 2y = 0 & \\ y(0) = 2, & y'(0) = 4\end{array}\right\} \tag{4.3.3}$$

Assume that the solution y and its derivative y' are of class E and that y' is continuous for $t \geq 0$. (By the definition of a solution in §1.2 y must be continuous for $t \geq 0$.) Then the second derivative y'' is continuous on the interval $(0, \infty)$ by (4.3.3), and so Theorem 4.2.2 can be applied to both y and y'. By the linearity (4.2.1) and by (4.2.3) it follows that

$$s^2Y(s) - 2s - 4 - 2(Y(s) - 2) + 2Y(s) = 0$$

$$Y(s) = \frac{2s}{s^2 - 2s + 2}$$

This function cannot be found in Table 4.4.1, but we can rewrite it by completing the square

$$\begin{aligned} Y(s) &= \frac{2s}{(s-1)^2+1} \\ &= 2\frac{s-1}{(s-1)^2+1} + 2\frac{1}{(s-1)^2+1} \end{aligned}$$

and hence by Table 4.4.1 numbers 11 and 12, we obtain

$$y(t) = 2e^t \cos t + 2e^t \sin t$$

We have thus shown that the boundary value problem (4.3.3) has a solution. We do not know yet if this is the only solution - this will be shown later in §5.2.

Example 3

$$\left.\begin{array}{ll}\dfrac{d^2y}{dt^2} + 4y = 8\sin 2t & \\ y(0) = 9, & y'(0) = 0\end{array}\right\} \tag{4.3.4}$$

Assume that the solution y and its derivative y' are of class E and that y' is continuous for $t \geq 0$. Then the second derivative y'' is continuous for $t > 0$ by (4.3.4) and Theorem 4.2.2 can be applied to both y and y'. By the linearity (4.2.1) and by (4.2.3) it follows that

$$s^2Y(s) - 9s + 4Y(s) = \frac{16}{s^2+4}$$

where Table 4.4.1, number 7, was used to transform the right hand side. Now solve for Y in the s-domain:

$$Y(s) = \frac{9s}{s^2+4} + \frac{16}{(s^2+4)^2}$$

and use Table 4.4.1, numbers 8 and 13, to read off the inverse Laplace transforms:

$$y(t) = 9\cos 2t + \sin 2t - 2t\cos 2t$$

As in the case of Example 2, the uniqueness will be discussed in §5.2.

- Do: Exercise 18 in §4.5.

Example 4

$$\left.\begin{aligned} &\frac{d^2y}{dt^2} + 4y = 8\sin 2t \\ &y(0) = 9, \qquad y(\tfrac{\pi}{4}) = 1 \end{aligned}\right\} \tag{4.3.5}$$

This example differs from the preceding one only in the second boundary condition. Where the boundary values in Example 3 were both given at the initial point $t = 0$, this example has boundary values at two different points. Thus Example 4 is called a *two-point boundary value problem.* We shall see in the next two examples that unexpected things can happen in two-point boundary value problems. Under the same assumptions as in Example 3, we obtain

$$s^2Y(s) - 9s - y'(0) + 4Y(s) = \frac{16}{s^2+4}$$

$$Y(s) = \frac{9s + y'(0)}{s^2+4} + \frac{16}{(s^2+4)^2}$$

By Table 4.4.1, numbers 7, 8, and 13, it follows that

$$y(t) = 9\cos 2t + \frac{1}{2}y'(0)\sin 2t + \sin 2t - 2t\cos 2t \tag{4.3.6}$$

in which the unknown constant $y'(0)$ still appears. The second boundary condition in (4.3.5) can now be applied:

$$1 = 0 + \frac{1}{2}y'(0) + 1 - 0$$

which implies that $y'(0) = 0$. Now we have the same initial values as in (4.3.4), so that the same answer must be obtained.

Example 5

$$\left.\begin{aligned} &\frac{d^2y}{dt^2} + 4y = 8\sin 2t \\ &y(0) = 9, \qquad y(\tfrac{\pi}{2}) = 1 \end{aligned}\right\} \tag{4.3.7}$$

This example differs from (4.3.5) only in that the second boundary condition in (4.3.5) is slightly changed, but it makes a profound difference in the solution. Proceeding as in Example 4, we obtain (4.3.6) as before. When $t = \frac{\pi}{2}$ is substituted in (4.3.6), we find that

$$1 = -9 + 0 + 0 + \pi$$

which is false. This means that it is impossible to find a function $y(t)$ which satisfies (4.3.7), and such that y and y' are continuous for $t \geq 0$ and of class E. In other words, the solution to the boundary value problem (4.3.7) does not exist.

Example 6

$$\left.\begin{aligned} &\frac{d^2y}{dt^2} + 4y = 8\sin 2t \\ &y(0) = 9, \qquad y(\tfrac{\pi}{2}) = \pi - 9 \end{aligned}\right\} \tag{4.3.8}$$

Again, as in Example 5, this example differs from (4.3.5) only in that the second boundary condition in (4.3.5) is changed. But in this case when $t = \frac{\pi}{2}$ is substituted in (4.3.6), we obtain an identity. This means that (4.3.6) is a solution to the boundary value problem (4.3.8), irrespective of the value chosen for $y'(0)$. Hence the solution is not unique - in fact, there are infinitely many solutions corresponding to different values for $y'(0)$.

- Do: Exercises 19, 20 in §4.5.

Note that the differential equations in the examples above are linear and the coefficients of y and the derivatives of y are constants. If the differential equation is nonlinear, then the transform of the nonlinear term is a problem. For example if y^2 appears in the differential equation, then $L[y^2]$ cannot be written in terms of $L[y]$ because we have no general rule by which the transform of a product of two functions can be written as a relation between the two transforms of the functions. Hence, the method of Laplace transforms can only be used for linear differential equations. Secondly, if the coefficient of y is not a constant, there is also no general rule to write $L[f(t)y]$ in terms of $L[y]$. In some special cases, however, Theorem 4.2.3 can be used, as in the following example.

Example 7

$$\left.\begin{array}{ll} \dfrac{d^2y}{dt^2} + t\dfrac{dy}{dt} - y = 0 & \\ y(0) = 0, & y'(0) = 4 \end{array}\right\} \tag{4.3.9}$$

Assume as before that the solution y and its derivative y' are of class E, and that y' is continuous for $t \geq 0$. Then the second derivative is continuous for $t > 0$ by (4.3.9) and Theorem 4.2.2 can be applied.

$$s^2 Y(s) - 4 - L[ty'] - Y(s) = 0$$

By (4.2.5) and (4.2.2) we find

$$L[ty'] = -\frac{d}{ds}(sY(s) - 0)$$

which then leads to a linear differential equation of the first order in the s-domain

$$\frac{dY}{ds} - \frac{s^2-2}{s}Y = -\frac{4}{s} \tag{4.3.10}$$

With the aid of the integrating factor

$$e^{-\int \frac{s^2-2}{s}\,ds} = e^{-\frac{1}{2}s^2 + 2\ln s} = s^2 e^{-\frac{1}{2}s^2}$$

(or by using Laplace transforms again on this new differential equation) we obtain

$$Y(s) = \frac{1}{s^2}\left(4 + ce^{\frac{1}{2}s^2}\right) \tag{4.3.11}$$

where c is an arbitrary constant. (Note that there is no boundary condition.) However, by Theorem 4.2.1 $Y(s)$ must satisfy the condition

$$\lim_{s\to\infty} Y(s) = 0$$

By the rule of L'Hôpital (see [64] page 264 or [16] page 464)

$$\lim_{s\to\infty} s^2 e^{-\frac{1}{2}s^2} = \lim_{s\to\infty} \frac{2s}{se^{\frac{1}{2}s^2}} = 0$$

and hence, the second term of (4.3.11) is unbounded as $s \to \infty$ unless $c = 0$. Thus the solution of (4.3.10) is

$$Y(s) = \frac{4}{s^2}$$

and by Table 4.4.1, number 1, we obtain the solution to (4.3.9):

$$y(t) = 4t$$

As will be seen in §5.2, this is the only solution to the initial value problem (4.3.9).

- Do: Exercise 21 in §4.5.

In many applications the resulting mathematical model is stated in terms of an integral equation, rather than a differential equation. Although we shall not construct such models in this book, it is interesting to note that some linear integral equations can also be handled by Laplace transforms, as the last example shows.

Example 8

$$\int_0^t y(\tau)\sin(t-\tau)\,d\tau = y(t) - 6t \tag{4.3.12}$$

Assume that y is of class E. By (4.2.7) we have

$$y(t) * \sin t = y(t) - 6$$

which can be transformed to the s-domain by (4.2.8):

$$Y(s)\frac{1}{s^2+1} = Y(s) - \frac{6}{s^2}$$

where Table 4.4.1, numbers 1 and 7, were used. After simplification

$$Y(s) = 6\frac{s^2+1}{s^4} = \frac{6}{s^2} + \frac{6}{s^4}$$

it follows immediately by Table 4.4.1, number 1, that

$$y(t) = 6t + t^3$$

Hence, the method to obtain the function which satisfies (4.3.12) by Laplace transforms is very similar to the method used in the previous examples.

- Do: Exercises 22, 23 in §4.5.
- Do: Project A in §4.6.

4.4 Table of Laplace Transforms

The usefulness of logarithms depends, to a large extent, on the availability of the logarithm of every real number, either in a table or in the memory of a computer. In the same way, a table of Laplace transforms provides a quick method to obtain a solution to a linear boundary value problem. It is, of course, an impossible task to show the Laplace transform of every transformable function in a table. In Table 4.4.1 some common functions which appear regularly in problems are included. Some books contain more extensive tables, for example [15] page 459, where the Laplace transforms of 121 functions are listed.

The properties of Laplace transforms such as (4.2.1) and Theorems 4.2.3 and 4.2.4 must, of course, be used to obtain the Laplace transforms of functions not in the table, as was shown in the examples of §4.3. Partial fractions also play an important part, as we had seen. In Table 4.4.1, numbers 5 and 6, the gamma function $\Gamma(k)$ is used. It is defined as the following integral:

$$\Gamma(k) = \int_0^\infty x^{k-1} e^{-x} \, dx \tag{4.4.1}$$

where k is a positive real number. The gamma function is an extension of the notion of

$$k! = 1 \times 2 \times 3 \times \ldots \times k$$

because if k in (4.4.1) is a positive integer, then

$$\Gamma(k) = (k - l)! \tag{4.4.2}$$

This follows immediately by induction with the aid of the property

$$\Gamma(k+1) = k\Gamma(k) \tag{4.4.3}$$

which is valid for any real number k (see Exercise 24 in §4.5). Another interesting fact is that (see [64] page 371)

$$\Gamma(\frac{1}{2}) = \sqrt{\pi} \tag{4.4.4}$$

The gamma function is tabulated for values between 0 and 1. The relation (4.4.3) is used to find the other values of $\Gamma(k)$.

- Do: Exercises 24, 25 in §4.5.

Number	$F(s) = \int_0^\infty e^{-st} f(t)\, dt$	$f(t)$
1	$\frac{n!}{s^{n+1}}$	$t^n \quad (n = 0, 1, 2, \ldots)$
2	$\frac{1}{s-a}$	e^{at}
3	$\frac{n!}{(s+a)^{n+1}}$	$t^n e^{-at} \quad (n = 0, 1, 2, \ldots)$
4	$\frac{1}{2}\sqrt{\pi} s^{-\frac{3}{2}}$	$\sqrt{t}$
5	$\frac{\Gamma(k+1)}{s^{k+1}}$	$t^k \quad (k > -1)$
6	$\frac{\Gamma(k+1)}{(s+a)^{k+1}}$	$t^k e^{-at} \quad (k > -1)$
7	$\frac{a}{s^2+a^2}$	$\sin at$
8	$\frac{s}{s^2+a^2}$	$\cos at$
9	$\frac{a}{s^2-a^2}$	$\sinh at$
10	$\frac{s}{s^2-a^2}$	$\cosh at$
11	$\frac{b}{(s-a)^2+b^2}$	$e^{at} \sin bt$
12	$\frac{s-a}{(s-a)^2+b^2}$	$e^{at} \cos bt$
13	$\frac{2a^3}{(s^2+a^2)^2}$	$\sin at - at \cos at$
14	$\frac{2as}{(s^2+a^2)^2}$	$t \sin at$
15	$\frac{s^2-a^2}{(s^2+a^2)^2}$	$t \cos at$
16	$\frac{s^3}{(s^2+a^2)^2}$	$\cos at - \frac{1}{2} at \sin at$

Table 4.4.1: **Laplace transforms**

4.5 Exercises

(1) Show that the transformation T of (4.1.2) is a linear transformation.

(2) If $f(t) = e^{kt}$ $(0 \le t < \infty)$ with k a constant, deduce $L[f]$ from the definition (4.1.3) and determine for which values of s the transform will exist.

(3) Repeat Exercise 2 with $f(t) = t^2$.

(4) Use the property of linearity (4.2.1) and Exercise 2 to find $L[\sinh kt]$ with k a constant. For which values of s does this transform exist?

(5) Given that

$$\int_0^\infty e^{-x^2}\, dx = \frac{1}{2}\sqrt{\pi}$$

use the definition (4.1.3) to find the Laplace transform of the function $f(t) = t^{-\frac{1}{2}}$.

(6) State, with reasons, whether the following functions are piecewise continuous on the given interval or not:

(a) $|t|$ $(-1 \le t \le 1)$

(b) $\ln |t|$ $(-1 \le t \le 1)$

(c) $\frac{t}{t-2}$ $(-1 \le t \le 1)$

(d) $\frac{t}{t-2}$ $(0 \le t \le 2)$

(e) $f(t) = \left\{ \begin{array}{ll} 2t + 3 & (0 \le t \le 1) \\ 2t - 3 & (-1 \le t < 0) \end{array} \right\}$

(7) Show that any polynomial is of exponential order α with α an arbitrary small positive number.

(8) Show that $f(t) = e^{at} \sin bt$ is of exponential order a.

(9) Suppose f and g are of exponential order. Show that $f + g$ and fg are also of exponential order.

(10) Investigate whether the following functions are of class E:

(a) $f(t) = t^2 e^{3t}$ $(t \ge 0)$

(b) $f(t) = \left\{ \begin{array}{ll} \frac{1}{t+2} & (0 \le t < 1) \\ 2 & (t > 0) \end{array} \right.$

(c) $f(t) = \begin{cases} e^{t^2} & (0 \le t < 3) \\ 0 & (t = 3) \\ -e^{3t} & (t > 3) \end{cases}$

(d) $f(t) = \begin{cases} t+4 & (0 \le t < 2) \\ 3t & (t \ge 2) \end{cases}$

(e) $f(t) = t^{-\frac{1}{2}} \quad (t \ge 0)$

(11) Show that if f and g are of class E, then $\alpha f + \beta g$ is also of class E with α and β arbitrary real numbers.

(12) Show that if f is of class E, then $\int_0^t f(\sigma)\, d\sigma$ is

(a) continuous on any finite interval and

(b) of class E.

(13) Prove Theorem 4.2.2 for the case when f' is piecewise continuous.

(14) Use (4.2.5) and Exercise 2 to obtain $L[t^2 e^{kt}]$ with k a constant.

(15) Use (4.2.6) and Exercise 3 to obtain $L[t^2 e^{kt}]$ with k a constant. Compare your answer with the answer of Exercise 14.

(16) Calculate the convolution of f and g when

(a) $f(t) = g(t) = 1 \quad (t \ge 0)$

(b) $f(t) = t^2, \quad g(t) = \cos t \quad (t \ge 0)$

(17) Use convolution and Table 4.4.1 to find

(a) $L^{-1}[\frac{1}{s^2(s-a)}]$

(b) $L^{-1}[\frac{9}{(s-1)^2(s-4)}]$

(18) Solve the following initial value problems by Laplace transforms:

(a) $y'' + y = 2, \quad y(0) = 4, \quad y'(0) = -2$

(b) $y'' + 5y' + 6y = e^{7t}, \quad y(0) = 0, \quad y'(0) = 2$

(c) $y'' + y = 5e^t \sin 2t, \quad y(0) = 3, \quad y'(0) = 1$

(d) $y'' + 4y' + 13y = 20e^{-t}, \quad y(0) = 1, \quad y'(0) = 3$

(19) Solve the following two-point boundary value problems by Laplace transforms:

(a) $y'' + 2y' + y = 0, \quad y(0) = 0, \quad y(1) = 1$

(b) $y'' + y = t, \quad y(0) = \pi, \quad y(\pi) = 0$

(c) $y'' + y = t, \quad y(\pi) = 0, \quad y'(0) = 1$

(d) $y'' + y = t, \quad y(0) = \pi, \quad y(\pi) = 1$

(20) Solve by Laplace transforms

$$\begin{cases} y''' - 2y'' + 5y' = 0 \\ y(0) = 0, \quad y'(0) = 1, \quad y(\pi) = 1 \end{cases}$$

(21) Use (4.2.5) to solve

(a) $y'' + \frac{1}{2}ty' - y = 1, \quad y(0) = y'(0) = 0$

(b) $ty'' + (t-1)y' + y = 0, \quad y(0) = y'(0) = 0$

Note that the zero function is a trivial solution of (b).

(22) Solve the followlng problem in two ways:

$$y'(t) + \int_0^t y(\tau)\, d\tau = 0, \quad y(0) = 1$$

(a) Differentiate the equation and solve the resulting second order differential equation. (Note that this is the equation for a simple harmonic motion.)

(b) Use Laplace transforms directly on the equation.

(23) Solve for $y(t)$ if:

(a) $y(t) = 2\sin t - 2\int_0^t y(\tau)\cos(t-\tau)\, d\tau$

(b) $y(t) = 2\sin t - 3\int_0^t y(\tau)\sin(t-\tau)\, d\tau$

(c) $y'(t) + y(t) = 1 - \int_0^t e^{t-\tau} y(\tau)\, d\tau, \quad y(0) = 0$

(24) (a) Prove (4.4.3) directly from the definition (4.4.1).

(b) Show that $\Gamma(1) = 1$.

(c) Use (4.4.3) to show that (4.4.2) holds when k is a positive integer.

(25) Use (4.4.3) and (4.4.4) to calculate $\Gamma(3.5)$.

4.6 Projects

Laplace transforms are especially useful in mathematical models where either a system of linear differential equations or second order differential equations must be solved. Since these type of models will be discussed in Chapters 5 and 6, respectively, we shall include here only one typical example in which a first order linear differential equation appears.

Project A

The mountain reedbuck (*redunca fulvorufula*) is a small deer common in the nature reserves of southern Africa. The owner of a private game reserve in the Karoo has to cull a number of his stock of reedbuck to prevent overgrazing. He is, however, not sure when and how many reedbuck must be removed, and is furthermore worried about the long-term effect of such a culling on his stock. So he approaches you as an expert on biological models to model this situation mathematically so that different strategies can be compared. You are given the following table of observed populations:

Year	Population
1988	260
1989	370
1990	500
1991	680
1992	950

(1) Construct a Malthus model of this population. (Use the method of least squares of §2.3 to calculate the parameters.)

(2) As in §2.5, modify the model in (1) to include a culling of $E(t)$ reedbucks per year, starting in 1992. Use Laplace transforms to solve this model by implementing a convolution to write $E(t)$ in terms of an integral in the t-domain.

(3) Consider the strategy of culling a fixed number of reedbuck every year. In the solution obtained in (2), let $E(t) = q$ where q is a constant. Show that there is a critical value $\bar{q}$ in the sense that if $q > \bar{q}$, then the population will be extinct in a finite time period, and if $q < \bar{q}$, then the population will increase. Calculate $\bar{q}$. What happens when $q = \bar{q}$? Draw a rough graph for the cases when q is

less than, equal to, and bigger than $\bar{q}$. Justify whether you would recommend this strategy to your client or not.

(4) Consider the strategy where a constant number q reedbucks are culled every year over a period of three years, after which the culling is stopped. Substitute this choice for $E(t)$ in the solution found in (2), to find the solution in this case. Is there again a critical value $\bar{q}$ as in (3)? Determine the value of q such that the reedbuck population will be back to the level of 1992 in 1998. Calculate the minimum population for this value of q during the years 1992 to 1998.

(5) Consider the strategy to cull q reedbucks every alternate year in such a way that the population is back to the original level at the end of the next year. Find the solution as in (4). Draw a rough graph to show the fluctuation of the population in this case. How does this strategy compare to the strategy in (4)? Which one will you recommend to your client? State the reasons for your choice.

4.7 Mathematical Background

This section contains the proofs of the theorems in Chapter 4. To prove that the operator L is one-to-one, a simple lemma is needed. The proof of this lemma rests on a famous theorem of Weierstrass:

Theorem

For any function $h(x)$, continuous on the closed interval $[a, b]$, and any positive number ϵ, there exists a polynomial $p(x)$ such that

$$|h(x) - p(x)| < \epsilon \quad \text{for all } x \in [a, b].$$

By this theorem any continuous function can be approximated as closely as you want, by a polynomial. Thus the theorem states that the set of all polynomials is dense in the set of continuous functions in the same sense that the set of rational numbers is dense in the set of real numbers. We shall not prove this theorem - a proof can be found in [15] page 426.

Lemma

If $h(x)$ is continuous on the closed interval $[0,1]$ and

$$\int_0^1 x^n h(x)\, dx = 0$$

for $n = 0, 1, 2, \ldots$ then $h(x) = 0$ on the interval $[0,1]$.

Proof

Let ϵ be an arbitrary small number. Then by the Weierstrass approximation theorem, there exists a polynomial $p(x)$ such that

$$|h(x) - p(x)| < \epsilon \qquad \text{for all } x \in [0,1].$$

Morever, by the hypothesis it follows that

$$\int_0^1 h(x)p(x)\, dx = 0$$

Hence

$$\begin{aligned} \int_0^1 h^2(x)\, dx &= \int_0^1 h(x)[h(x) - p(x)]\, dx \\ &\leq \int_0^1 |h(x)|\, |h(x) - p(x)|\, dx \\ &< \epsilon \int_0^1 |h(x)|\, dx \end{aligned}$$

Since ϵ was arbitrary it follows that

$$\int_0^1 h^2(x)\, dx = 0$$

If $h(\xi) \neq 0$ at some $\xi \in [0,1]$, then by the continuity of $h(x)$, it is nonzero over a small interval containing the point $x = \xi$. Hence, the integral of h^2 over this small interval is nonzero, which leads to a contradiction.

Theorem 4.2.1

If f is of class E with exponential order α, then the Laplace transform of f exists for all $s > \alpha$. Moreover, $\lim_{s\to\infty} L[f] = 0$. Two functions of class E which are not equal have different Laplace transforms.

Proof

Since f is of class E, there exists a real number T such that

$$|f(t)| < ce^{\alpha t} \qquad \text{for all } t > T.$$

Since f is piecewise continuous on the closed interval $[0, T]$, the integral

$$\int_0^T e^{-st} f(t)\, dt$$

will exist - this follows immediately by writing the integral as the sum of a finite number of integrals, each with a continuous integrand. Moreover,

$$\left|e^{-st} f(t)\right| < ce^{-(s-\alpha)t} \qquad \text{for all } t > T$$

and so the integral

$$\left|\int_T^\infty e^{-st} f(t)\, dt\right| < c\int_T^\infty e^{-(s-\alpha)t}\, dt = \frac{1}{s-\alpha} e^{-(s-\alpha)T}$$

provided that $s > \alpha$. Hence, the Laplace transform of f exists for $s > \alpha$. Suppose that $L[f] = L[g]$. Let

$$h(t) = f(t) - g(t)$$

then $h(t)$ is also of class E with exponential order α, and $H(s) = L[h] = 0$ for $s > \alpha$. Define

$$\bar{h}(t) = \int_0^t e^{-\beta\tau} h(\tau)\, d\tau$$

where β is any number larger than α. Then $\bar{h}(t)$ is continuous (see Exercise 12 in §4.5), $\bar{h}(0) = 0$ and $\bar{h}(\infty) = H(\beta) = 0$. Moreover,

$$\begin{aligned} H(s+\beta) &= \int_0^\infty e^{-(s+\beta)t} h(t)\, dt \\ &= \int_0^\infty e^{-st} e^{-\beta t} h(t)\, dt \\ &= e^{-st}\bar{h}(t)\big|_0^\infty + s\int_0^\infty e^{-st}\bar{h}(t)\, dt \end{aligned}$$

by integration by parts. Since $H(s+\beta) = 0$ for all $s \geq 1$, it follows that

$$\int_0^\infty e^{-st}\bar{h}(t)\,dt = 0 \qquad \text{for all } s \geq 1$$

Let $z = e^{-t}$, then

$$\int_0^1 z^{s-1}\bar{h}(-\ln z)\,dz = 0 \qquad \text{for all } s \geq 1$$

Since $\bar{h}(-\ln z) = 0$ when $z = 0$ and $z = 1$, and thus continuous on the closed interval $[0, 1]$, the lemma implies that $\bar{h}(-\ln z) = 0$ for $z \in [0, 1]$, and hence $\bar{h}(t) = 0$ for $t \geq 0$. By the definition of $\bar{h}(t)$, it now follows that

$$\int_0^t e^{-\beta\tau}h(\tau)\,d\tau = 0 \qquad \text{for all } t \geq 0$$

But this in turn implies that

$$I(t) = \int_a^t e^{-\beta\tau}h(\tau)\,d\tau = 0$$

for any arbitrary number a with $a \geq 0$. Suppose that h is continuous on an interval (a, b). Since $I(t)$ is identically zero on this interval, the derivative of this integral is also zero on this interval, and hence $h(t)$ is zero on (a, b). Since $h(t)$ is piecewise continuous, we have proved that $h(t)$ is zero over every interval where $h(t)$ is continuous. There remain a finite number of discontinuities, but by the definition of a function of class E, $h(t)$ is also zero at these points of discontinuity, since $h(t)$ is zero on both sides of the discontinuity. Hence, it was shown that if $L[f] = L[g]$, then $f = g$.

Finally, since $f(t)$ is of class E,

$$\left|f(t)e^{-st}\right| < Me^{-(s-\alpha)t} \qquad \text{for } t \geq 0$$

with M a positive constant, where the boundedness of a piecewise continuous function on a finite interval was used. Hence, for $s > \alpha$,

$$\left|\int_0^\infty e^{-st}f(t)\,dt\right| < M\int_0^\infty e^{-(s-\alpha)t}\,dt = \frac{M}{s-\alpha}$$

which shows that $\lim_{s\to\infty} L[f] = 0$.

Theorem 4.2.2

If f is continuous on the interval $[0, \infty)$ and of exponential order α, and the derivative f' of f is continuous on the interval $(0, \infty)$, then the Laplace transform of f' exists for $s > \alpha$ and

$$L[f'] = sL[f] - f(0)$$

Proof

$$\begin{aligned} L[f'] &= \int_0^\infty e^{-st} f'(t)\, dt \\ &= \lim_{b \to \infty} \int_0^b e^{-st} f'(t)\, dt \\ &= \lim_{b \to \infty} \left[e^{-st} f(t)|_0^b + s \int_0^b e^{-st} f(t)\, dt \right] \\ &= \lim_{b \to \infty} e^{-st} f(b) - f(0) + sL[f] \end{aligned}$$

There exist constants c, T, and α such that for $t > T$

$$|f(t)| < ce^{\alpha t} \quad \text{and hence} \quad |e^{-st} f(t)| < ce^{(\alpha - s)t}$$

If $s > \alpha$, this implies that $\lim_{b \to \infty} e^{-st} f(t) = 0$. As in the proof of Theorem 4.2.1, the existence of $L[f]$ is guaranteed.

Theorem 4.2.3

If the function f is of class E with exponential order α, then so is the integral of f and for $s > \alpha$

$$L\left[\int_0^t f(\sigma)\,d\sigma\right] = \frac{1}{s}L[f] \qquad (4.7.1)$$

Moreover, all the derivatives of $L[f]$ exist, and for any positive integer n and $s > \alpha$

$$L[t^n f(t)] = (-1)^n \frac{d^n}{ds^n} L[f] \qquad (4.7.2)$$

For any real number a, $s > \alpha + a$, and $F(s) = L[f]$,

$$L[e^{at} f(t)] = F(s-a) \qquad (4.7.3)$$

Proof

Let $\bar{f}(t) = \int_0^t f(\sigma)\,d\sigma$. Then $\bar{f}$ is continuous for $t \geq 0$, differentiable with derivative f, and of class E. (See Exercise 12 in §4.5). By (4.2.2) we have

$$L[\bar{f}'] = sL[\bar{f}] - \bar{f}(0)$$

from which (4.7.1) immediately follows. To show (4.7.3) we need only the definition (4.1.3)

$$\begin{aligned} L[e^{at} f(t)] &= \int_0^\infty e^{-st} e^{at} f(t)\,dt \\ &= \int_0^\infty e^{-(s-a)t} f(t)\,dt \\ &= F(s-a) \end{aligned}$$

Finally, to prove (4.7.2), note that $t^n f(t)$ is also of class E for $n = 0, 1, 2, \ldots$ (see Exercises 7 and 9 in §4.5) with exponential order $\alpha + \epsilon$ where ϵ is arbitrarily small. Hence $L[t^n f(t)]$ exists. If the interchange between the integral and the derivative with respect to s is permissible,

(4.7.2) immediately follows by induction:

$$\begin{aligned}
\frac{d^n}{ds^n}L[f] &= \frac{d}{ds}\left(\frac{d^{n-1}}{ds^{n-1}}L[f]\right) \\
&= (-1)^{n-1}\frac{d}{ds}L[t^{n-1}f(t)] \\
&= (-1)^{n-1}\int_0^\infty \frac{\partial}{\partial s}\left(e^{-st}\right)t^{n-1}f(t)\,dt \\
&= (-1)^n\int_0^\infty e^{-st}t^n f(t)\,dt \\
&= (-1)^n L[t^n f(t)]
\end{aligned}$$

This interchange is permissible if $L[t^n f(t)]$ converges uniformly with respect to s (see [15] p. 42 or [35] p. 315). The uniform convergence follows immediately by the Weierstras M-test (see [35] p. 245):

$$\left|t^n f(t)e^{-st}\right| < Ke^{-(s-\alpha-\epsilon)t}$$

for $t \geq 0$ with K a positive constant, where the boundedness of a piecewise continuous function on a finite interval was used. Moreover, if $s \geq \alpha_0$ where $\alpha_0 > \alpha + \epsilon$, then we have

$$\left|t^n f(t)e^{-st}\right| < Ke^{-(\alpha_0-\alpha-\epsilon)t} = M(t)$$

for $t \geq 0$. This function $M(t)$ is integrable from zero to infinity and is independent of s. Hence, the uniform convergence of $L[t^n f(t)]$ follows for $s \geq \alpha_0$. This completes the proof.

Theorem 4.2.4

If f, g, and h are any piecewise continuous functions on a closed interval $[0, T]$, then $f*g(t)$ is continuous on $[0, T]$, and

(i) $f * g = g * f$
(ii) $f * (\alpha g) = \alpha(f * g)$ for any real number α
(iii) $(f * g) * h = f * (g * h)$
(iv) $f * (g + h) = f * g + f * h$

If f and g are of class E with exponential order α, then for $s > \alpha$

$$L[f * g] = L[f]L[g] \tag{4.7.4}$$

Proof

The proof that $f * g(t)$ is continuous is the same as Exercise 12 in §4.5. The property (i) of commutativity follows immediately when a new variable of integration $u = t - \tau$ is introduced in the convolution integral (4.2.7). The proof of properties (ii) and (iv) is a direct consequence of the properties of the definite integral.

To prove (iii) we need only to reverse the order of integration. By the definition (4.2.7) we have

$$f * (g * h) = f * (h * g) = \int_0^t f(\tau) \int_0^{t-\tau} h(\sigma) g(t - \tau - \sigma)\, d\sigma\, d\tau$$

which is an iterated integral over the triangular region shown in Figure 4.7.1.

On the other hand we also have by the definition (4.2.7) that

$$(f * g) * h = h * (f * g) = \int_0^t h(\sigma) \int_0^{t-\sigma} f(\tau) g(t - \sigma - \tau)\, d\tau\, d\sigma$$

which is an iterated integral over the triangular region shown in Figure 4.7.2. Since the integrands and the triangular regions are the same, the property (iii) follows.

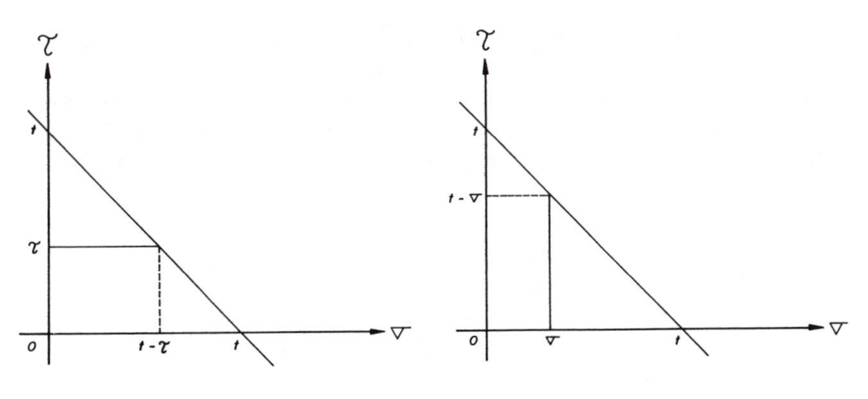

Figure 4.7.1: **Integration w.r.t σ first**

Figure 4.7.2: **Integration w.r.t τ first**

Finally we prove the property (4.7.4). For a fixed non-negative number τ we have for $s > \alpha$

$$\begin{aligned}\int_0^\infty e^{-st} g(t-\tau)\,d\tau &= \int_{-\tau}^\infty e^{-s(\sigma+\tau)} g(\sigma)\,d\sigma \\ &= e^{-s\tau}\int_0^\infty e^{-s\sigma} g(\sigma)\,d\sigma \\ &= e^{-s\tau} L[g]\end{aligned}$$

where $g(t)$ was defined to vanish for all negative values of t as in (4.1.4). Hence

$$\begin{aligned}L[f]L[g] &= \int_0^\infty e^{-s\tau} f(\tau) L[g]\,d\tau \\ &= \int_0^\infty f(\tau)\int_0^\infty e^{-st} g(t-\tau)\,dt\,d\tau \\ &= \lim_{T\to\infty}\int_0^T\int_0^\infty f(\tau)e^{-st} g(t-\tau)\,dt\,d\tau\end{aligned}$$

The interchange of the order of integration of the definite and the improper integrals (for a fixed T) is valid provided that the improper integral converges uniformly with respect to the parameter τ (see [15] p. 42). As before, the uniform convergence follows by the Weierstrass M-test (see [35] p. 245):

$$\left|f(\tau)e^{st} g(t-\tau)\right| < Ke^{\alpha t}e^{-st}e^{\alpha(t-\tau)} = Ke^{-(s-\sigma)t} \tag{4.7.5}$$

where the exponential order of f and g were used. Since $s > \alpha$, we have a function $M(t)$ which is integrable from zero to infinity and which is independent of τ, from which the uniform convergence follows. We thus have

$$\begin{aligned}L[f]L[g] &= \lim_{T\to\infty}\int_0^\infty\int_0^T f(\tau)e^{-st} g(t-\tau)\,d\tau\,dt \\ &= \lim_{T\to\infty}\left\{\int_0^T\int_0^T f(\tau)e^{-st} g(t-\tau)\,d\tau\,dt\right. \\ &\qquad \left. + \int_T^\infty\int_0^T f(\tau)e^{-st} g(t-\tau)\,d\tau\,dt\right\}\end{aligned}$$

By (4.7.5) the absolute value of the second integral is less than

$$K\int_T^\infty\int_0^T e^{-(s-\alpha)t}\,d\tau\,dt = \frac{KT}{s-\alpha}e^{-(s-\alpha)T}$$

which tends to zero as $T \to \infty$. In the first integral $g(t - \tau) = 0$ when $\tau > t$ (see the convention that a function is zero on the negative axis at the end of §4.1). Hence

$$\begin{aligned} L[f]L[g] &= \lim_{T \to \infty} \int_0^T e^{-st} \int_0^t f(\tau)g(t-\tau)\, d\tau\, dt \\ &= L[f * g] \end{aligned}$$

which completes the proof.

5

Simultaneous Linear First Order Differential Equations

5.1 Introduction

In §2.6 where we modelled the optimization of profit, we encountered the situation of two differential equations acting simultaneously in equations (2.6.1) and (2.6.2):

$$\begin{aligned} \frac{du}{dt} &= kau, & u(0) = \alpha \\ \frac{dw}{dt} &= a(1-k)u, & w(0) = 0 \end{aligned}$$

Fortunately we could solve these equations easily by solving the first equation first, and then by substituting in the second equation, because the variable w is absent in the first equation.

In many mathematical models we obtain a system of first order differential equations of the form

$$\left.\begin{aligned} \frac{du_1}{dt} &= F_1(u_1, u_2, , u_n, t) \\ \vdots & \quad \vdots \\ \frac{du_n}{dt} &= F_n(u_1, u_2, , u_n, t) \end{aligned}\right\} \tag{5.1.1}$$

where the $F_i, i = 1, 2, \ldots, n$ are given functions of the $n+1$ real variables $u_1, u_2, \ldots, u_n, t$. We say that the system is then in *normal form.* If we define the vectors

$$\boldsymbol{u} = \begin{bmatrix} u_1 \\ u_2 \\ \vdots \\ u_n \end{bmatrix} \qquad \text{and} \qquad \boldsymbol{F} = \begin{bmatrix} F_1 \\ F_1 \\ \vdots \\ F_1 \end{bmatrix}$$

then (5.1.1) can be written in a more compact form

$$\frac{d\boldsymbol{u}}{dt} = \boldsymbol{F}(\boldsymbol{u}, t). \tag{5.1.2}$$

Given the boundary conditions of the form

$$\boldsymbol{u}(a) = \boldsymbol{\alpha} \tag{5.1.3}$$

the question is whether a solution exists. First, let us be precise about the word "solution" for a system of equations.

Definition 5.1.1

A solution of a given system of boundary value problems on the interval I is a vector function of which each component is continuous on I, which satisfies the system of differential equations at every point $t \in i(I)$ and agrees with the prescribed boundary values.

Note that $n = 1$ in (5.1.1) is the case of a single differential equation, and then this definition is the same as the original definition in §1.2.

By extending the definition of a Lipschitz condition to vector functions, the existence and uniqueness theorems of §3.2 can be shown to hold also for a normal system of differential equations (5.1.2) with boundary conditions (5.1.3). The proof is very similar to the proof of Theorem 3.2.1 with obvious modifications where scalars are replaced by vectors (see, for example, [9] Chapter V). We shall return to systems like (5.1.2), (5.1.3) in Chapter 7.

In this chapter we shall concentrate on models which utilize a special case of (5.1.2), namely a system of *linear first order differential equations*. Since we shall find a solution explicitly in each model, we shall only concern ourselves here with a general theorem on uniqueness.

Consider a system of n linear first order differential equations:

$$\left.\begin{aligned}
\frac{du_1}{dt} &= b_{11}(t)u_1 + b_{12}(t)u_2 + \ldots + b_{1n}(t)u_n + f_1(t) \\
\frac{du_2}{dt} &= b_{21}(t)u_1 + b_{22}(t)u_2 + \ldots + b_{2n}(t)u_n + f_2(t) \\
&\vdots \\
\frac{du_n}{dt} &= b_{n1}(t)u_1 + b_{n2}(t)u_2 + \ldots + b_{nn}(t)u_n + f_n(t)
\end{aligned}\right\} \tag{5.1.4}$$

on the interval $[a, T]$ with the same boundary values as in (5.1.3)

$$\left.\begin{array}{rcl} u_1(a) & = & \alpha_1 \\ u_2(a) & = & \alpha_2 \\ \vdots & & \vdots \\ u_n(a) & = & \alpha_n \end{array}\right\} \tag{5.1.5}$$

where $b_{ij}(t)$ with $i, j = 1, 2, \ldots, n$ and $f_i(t)$ with $i = 1, 2, \ldots, n$ are given functions defined on $[a, T]$, and α_i with $i = 1, 2, \ldots, n$ are given constants.

As in (5.1.2) we can use vector notation to abbreviate (5.1.4) to

$$\frac{d\boldsymbol{u}}{dt} = B(t)\boldsymbol{u}(t) + \boldsymbol{f}(t)$$

where B is the matrix $(b_{ij}(t))_{n \times n}$ and $\boldsymbol{u}$ and $\boldsymbol{f}$ are column vectors with n components.

We show in §5.10 that the Gronwall lemma (see §2.14) can be used to prove the uniqueness of the solution of (5.1.4), (5.1.5) in much the same way as in Theorem 2.4.1.

Theorem 5.1.1

Suppose that each of the coefficients $b_{ij}(t)$ for $i, j = 1, 2, \ldots, n$ in (5.1.4) are continuous on the closed interval $[a, T]$. If $\boldsymbol{u}$ and $\boldsymbol{v}$ are solutions of the system (5.1.4), (5.1.5), then $\boldsymbol{u}(t) = \boldsymbol{v}(t)$ for all $t \in [a, T]$.

The general theory of systems of linear differential equations in terms of matrices and eigenvectors will not be discussed in this chapter. We shall instead look at a few typical simple examples to illustrate some modelling concepts and some ideas to solve these boundary value problems.

- Do: Exercises 1, 2 in §5.8.

5.2 Projectile Trajectories with Air Resistance

The influence of air resistance on the motion of a particle when falling in the gravitational field of the earth was discussed in §2.9, and air resistance also played a role in §2.10 and §2.11. In each case the motion was rectilinear so that the coordinate axis for the displacement was chosen in the direction of the motion. We shall now model the motion of a projectile which is launched at an acute angle to the vertical, so that the motion can be described in a plane. Let us briefly discuss each of the stages of the modelling process:

Identification: Given the initial velocity, the *trajectory* (or flight path) of the projectile must be found, as well as the *height* (that is, the maximal vertical distance) and the *range* (that is, the maximal horizontal distance) and the *flight time* (that is, the total time for the flight). The trajectory and the range are obviously important for military purposes, but the flight time can also be crucial; for example, if the projectile is a bomb, the timefuse is normally set to detonate at a prescribed altitude. The basic mechanism which determines the behaviour of the projectile is, of course, Newton's second law of mechanics.

Assumptions: As in §2.11 we shall make the two basic assumptions:

Assumption (L)

The air resistance is directly proportional to the speed of the projectile at any time during the flight, but in the opposite direction.

Assumption (N) *The height of the trajectory (maximum altitude) of the projectile is small enough so that the acceleration of gravity g remains constant during the flight.*

We shall also assume that the only forces acting on the projectile are the gravity of the earth and the air resistance. Thus the motion can be described in a vertical plane. We shall assume that the range of the trajectory is such that the surface of the earth can be considered to be flat, or at least that the launch pad and the point where the projectile lands are on the same horizontal level, with no obstructions in between in the flight path.

Construction: Choose Cartesian coordinates with the origin at the point where the projectile will be launched, the x-axis in the horizontal direction, and the y-axis vertically upward. Let the initial velocity of the projectile be $\boldsymbol{V}$ with magnitude V and θ the angle between $\boldsymbol{V}$ and the positive x-axis. Let $\boldsymbol{v}(t)$ denote the velocity at any time t with $t = 0$ when the motion starts. For convenience we shall use the notation f' for the derivative of f with respect to t. Then we can write

$$\boldsymbol{v}(t) = x'(t)\boldsymbol{i} + y'(t)\boldsymbol{j}$$

where $\boldsymbol{i}$ and $\boldsymbol{j}$ are the usual unit vectors in the direction of the positive x- and y-axes, respectively.

We can now utilize Newton's second law of mechanics (see (2.9.1) and the discussion in §2.9)

$$\text{mass} \times \text{acceleration} = \text{total external force}$$

and Assumptions (L) and (N), above, to obtain the following vector equation:

$$m\frac{d\boldsymbol{v}}{dt} = -k\boldsymbol{v} - mg\boldsymbol{j} \tag{5.2.1}$$

in which m denotes the mass of the projectile and k is a positive constant. Furthermore, the initial velocity

$$\boldsymbol{v}(0) = \boldsymbol{V} \tag{5.2.2}$$

is known, which completes the mathematical description of the velocity at any time t during the flight.

Analysis: If we separate the components of (5.2.1) and (5.2.2), we obtain a system of two simultaneous differential equations:

$$\left.\begin{aligned} \frac{dx'}{dt} &= -\frac{k}{m}x', & x'(0) &= V\cos\theta \\ \frac{dy'}{dt} &= -\frac{k}{m}y' - g, & y'(0) &= V\sin\theta \end{aligned}\right\} \tag{5.2.3}$$

Both these first order differential equations are separable (see (2.1.9)). Using the initial values we obtain the unique solutions

$$\left.\begin{aligned} x' &= V\cos\theta e^{-\frac{kt}{m}} \\ y' &= \tfrac{m}{k}\left[-g + \left(g + \tfrac{k}{m}V\sin\theta\right)e^{-\frac{kt}{m}}\right] \end{aligned}\right\} \qquad (5.2.4)$$

By the choice of axes we also have

$$x(0) = y(0) = 0$$

After integrating the equations (5.2.4) with respect to t, it follows that

$$\left.\begin{aligned} x(t) &= \tfrac{m}{k}V\cos\theta\left(1 - e^{-\frac{kt}{m}}\right) \\ y(t) &= -\tfrac{m}{k}gt + (\tfrac{m}{k})^2(g + \tfrac{k}{m}V\sin\theta)(1 - e^{-\frac{kt}{m}}) \end{aligned}\right\} \qquad (5.2.5)$$

Hence the displacement and velocity of the projectile at any time t after launching is known. We can now answer the questions posed in the Identification stage above about the motion of the projectile. For example, the trajectory can be found by eliminating t in the equations (5.2.5). By the first equation we have

$$t = -\frac{m}{k}\ln\left(1 - \frac{kx}{mV\cos\theta}\right)$$

which can be substituted in the second equation to obtain the trajectory of the projectile:

$$y = \frac{m^2 g}{k^2}\ln\left(1 - \frac{kx}{mV\cos\theta}\right) + \frac{x}{V\cos\theta}\left(\frac{mg}{k} + V\sin\theta\right) \qquad (5.2.6)$$

The height of the trajectory follows when the vertical speed is zero: $y' = 0$. By (5.2.4) this happens after a time t_1 where

$$t_1 = \frac{m}{k}\ln\left(1 + \frac{k}{mg}V\sin\theta\right)$$

and the coordinates (x_1, y_1) of the point of maximum altitude can then be obtained by substituting t_1 in (5.2.5):

$$\left.\begin{aligned} x_1 &= \frac{mV^2\sin\theta\cos\theta}{mg + kV\sin\theta} \\ y_1 &= -\frac{m^2 g}{k^2}\ln\left(1 + \frac{k}{mg}V\sin\theta\right) + \frac{m}{k}V\sin\theta \end{aligned}\right\} \qquad (5.2.7)$$

The range and flight time can be calculated from the equations above, but cannot be found explicitly (see Exercise 3 in §5.8 where an interesting relationship between these quantities must be proved).

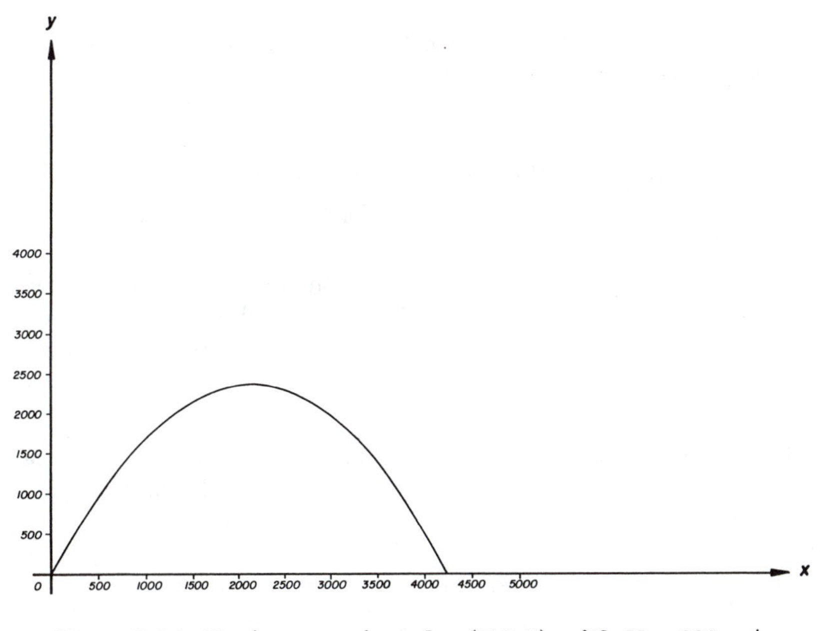

Figure 5.2.1: **Trajectory given by (5.2.6) with** $V = 200$ m/s, $\theta = 60°$, and $\frac{m}{k} = 450$ s

Interpretation: If the curve given by (5.2.6) is drawn for prescribed values of the parameters V, θ,and k, as in Figure 5.2.1, it is clear that the model provides an acceptable trajectory.

- Read: [37] page 26 or [36] page 23 where the case when air resistance is proportional to the square of the speed is discussed.
- Do: Exercises 3, 4, 5 in §5.8.
- Do: Projects A and B^* in §5.9.

5.3 Romantic Relationships

Life would be very dull without the excitement (and sometimes pain) of romance! Believe it or not, even love affairs can be modelled by differential equations. Steven Strogatz suggested a model for the following situation in [60]:

Suppose William is in love with Zelda, but Zelda is a fickle lover. The more William loves her, the more she dislikes him - but when he loses interest in her, her feelings for him warm up. On the other hand, William reacts to her: when she loves him, his love for her grows and when she loses interest, he also loses interest.

Let

$$\begin{aligned} w(t) &= \text{William's feelings for Zelda at time } t \\ z(t) &= \text{Zelda's feelings for William at time } t \end{aligned}$$

where positive and negative values of the variables w and z denote love and dislike, respectively. The exact units by which these variables can be measured will be left to the imagination of the reader. (Perhaps the love between Romeo and Juliet can be cannonized in the name of "romjul" for these units!)

We shall assume that the functions $w(t)$ and $z(t)$ are of class E, continuous and differentiable with respect to t for $t > 0$ with continuous first derivatives.

Assumption (P)

The change in one person's feelings in the time interval $[t, t + \delta t]$ is directly proportional to the other person's feelings over the same time interval.

Let w and z change to $w + \delta w$ and $z + \delta z$ respectively, in the time interval $[t, t + \delta t]$. Then by Assumption (P)

$$\begin{aligned} \delta w &= a \int_t^{t+\delta t} z(\tau)\, d\tau = az(t_1)\delta t \quad (t < t_1 < t + \delta t) \\ \delta z &= -b \int_t^{t+\delta t} w(\tau)\, d\tau = -bz(t_2)\delta t \quad (t < t_2 < t + \delta t) \end{aligned}$$

where a and b are positive constants. Note the minus sign to show that z decreases by δz when w is positive, and conversely, z increases by δz when w is negative. Divide by δt and let δt tend to zero:

$$\left.\begin{aligned} \frac{dw}{dt} &= az \qquad w(0) = \alpha \\ \frac{dz}{dt} &= -bw \quad z(0) = \beta \end{aligned}\right\} \tag{5.3.1}$$

where α and β denote the initial feelings of William and Zelda, respectively.

Having constructed the mathematical model (5.3.1), we now move into the Analysis stage. To see how the relationship between William and Zelda will develop, we must solve (5.3.1). We shall use Laplace transforms to do this. (Alternatively, the solution can also be found by substituting $w = Ae^{\lambda t}$, $z = Be^{\lambda t}$ in (5.3.1) to obtain an equation in λ called the characteristic equation - in this case $\lambda^2 + ab = 0$, so that both w and z are linear combinations of $\cos\sqrt{ab}t$ and $\sin\sqrt{ab}t$.)

Since both $w(t)$ and $z(t)$ are of exponential order, continuous and with continuous derivatives, it follows from Theorem 4.2.2 that the transforms

$$W(s) = L[w(t)] \quad \text{and} \quad Z(s) = L[z(t)]$$

exist. Then (5.3.1) is transformed to

$$\left.\begin{aligned} sW(s) - \alpha &= aZ(s) \\ sZ(s) - \beta &= -bW(s) \end{aligned}\right\} \tag{5.3.2}$$

which is a system of two linear simultaneous equations with the unique solution

$$\begin{aligned} W(s) &= \frac{\alpha s + a\beta}{s^2 + ab} \\ Z(s) &= \frac{\beta s - b\alpha}{s^2 + ab} \end{aligned}$$

To find the inverse transform, we use Table 4.4.1, numbers 7 and 8:

$$\left.\begin{aligned} w(t) &= \alpha\cos\sqrt{ab}t + \sqrt{\tfrac{a}{b}}\beta\sin\sqrt{ab}t \\ z(t) &= \beta\cos\sqrt{ab}t - \sqrt{\tfrac{b}{a}}\alpha\sin\sqrt{ab}t \end{aligned}\right\} \tag{5.3.3}$$

By Theorem 5.1.1 we know that this solution of (5.3.1) is unique. We can rewrite (5.3.3) in a more compact form by introducing a phase angle ϕ :

$$\tan\phi = \sqrt{\frac{a}{b}}\frac{\beta}{\alpha}$$

where $0 \le \phi < 360°$ and the quadrant is determined by the signs of α and β. (For example, $\alpha > 0$ and $\beta < 0$ implies that ϕ is in the fourth quadrant where $270° < \phi < 360°$.) Then (5.3.3) is

$$\left.\begin{aligned} w(t) &= \sqrt{\tfrac{b\alpha^2 + a\beta^2}{b}}\cos(\sqrt{ab}t - \phi) \\ z(t) &= -\sqrt{\tfrac{b\alpha^2 + a\beta^2}{a}}\sin(\sqrt{ab}t - \phi) \end{aligned}\right\} \tag{5.3.4}$$

The graphs of $w(t)$ and $z(t)$ are shown in Figure 5.3.1 for the case when $a > b$.

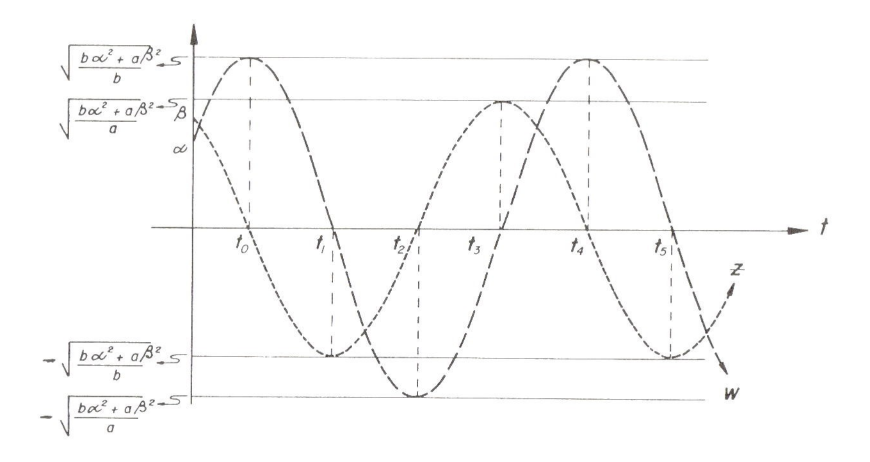

Figure 5.3.1: **Graphs of (5.3.4) with** $a > b$

At time $t_0 = \frac{\phi}{\sqrt{ab}}$, William's love for Zelda is a maximum

$$w(t_0) = \sqrt{\frac{b\alpha^2 + a\beta^2}{b}}$$

and Zelda is neutral in her attitude to William. As time moves on, Zelda starts to dislike William ($z(t) < 0$) and William's love also diminishes until it is zero at time $t_1 = \frac{2\phi+\pi}{2\sqrt{ab}}$ and Zelda's dislike

$$z(t_1) = -\sqrt{\frac{b\alpha^2 + a\beta^2}{a}}$$

is at its worst. But for $t > t_1$, William starts to dislike Zelda who is now looking with new eyes at William. At $t_2 = \frac{\phi+\pi}{\sqrt{ab}}$ Zelda has lost her dislike for William, but unfortunately William's dislike for Zelda is at its worst! For $t > t_2$ Zelda is falling in love with William, and consequently William is also losing his dislike for her until at $t_3 = \frac{2\phi+3\pi}{2\sqrt{ab}}$ William is over his dislike and Zelda is madly in love. But now, as William is falling in love, Zelda's love is cooling until at $t_4 = \frac{\phi+2\pi}{\sqrt{ab}}$ we have again the same situation as at $t = t_0$. Now the whole process repeats itself. Thus, William and Zelda are in a perpetual cycle of love and hate - in fact, they love each other simultaneously for only one quarter of the time!

We can also eliminate t in (5.3.4) by squaring and adding the two equations:

$$\frac{bw^2}{b\alpha^2 + a\beta^2} + \frac{az^2}{b\alpha^2 + a\beta^2} = 1 \qquad (5.3.5)$$

This represents an ellipse in the (z, w)-plane, as shown in Figure 5.3.2 where again $a > b$. In this figure we can think of the time t as a parameter which, by (5.3.4), defines points on this ellipse. We have shown, for example, the corresponding points for $t = t_0, t_1, t_2, t_3$, and t_4 in Figure 5.3.2, and it should be clear that as $t > t_4$, we shall again obtain the same points as we move anticlockwise on the ellipse.

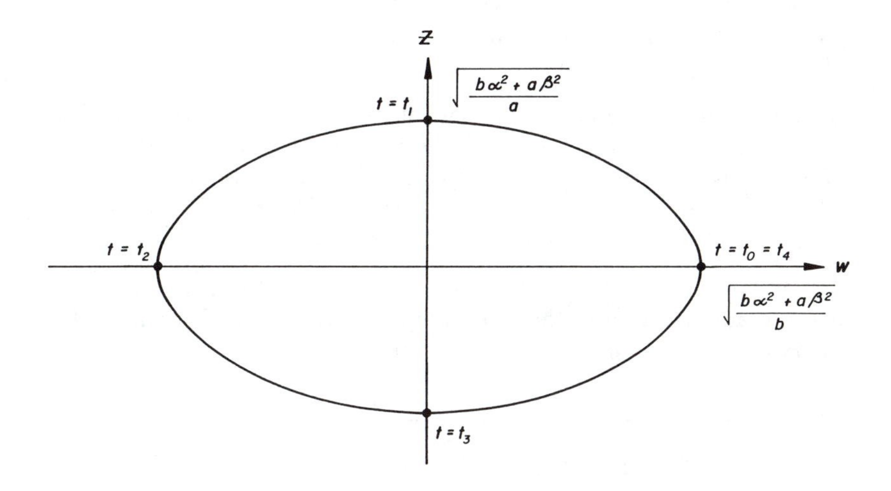

Figure 5.3.2: **Graph of (5.3.5) in the phase plane**

In a problem like (5.3.1) where two dependent variables w and z are given in terms of an independent variable t, we can either look at the graphs of $w(t)$ and $z(t)$, as in Figure 5.3.1, or eliminate the independent variable (if possible) to obtain a relation between w and z, as shown in Figure 5.3.2. The plane whose points can be identified with coordinates on the (z, w)-axes is called the *phase plane*. Note that if t does not appear explicitly on the right hand sides of (5.3.1), then the two equations can be written as a single equation

$$\frac{dz}{dw} = -\frac{bw}{az} \qquad (5.3.6)$$

which can then be solved (see Exercise 6 in §5.8). We call this type of differential equation an *autonomous system*. We shall study autonomous systems in Chapter 7.

- Do: Exercises 6, 7 in §5.8.
- Do: Project C in §5.9.

5.4 Neutron Flow

In this section we look at a simplified representation of a nuclear reaction as neutrons pass through a rod. We shall consider the rod as a one dimensional object and model it as the interval $[0, a]$ on the x-axis. As a neutron passes through the rod, it may interact with an atom in the rod with the result that the original neutron is replaced by two neutrons moving in the same direction and one neutron moving in the opposite direction.

Suppose that a given stream of neutrons enters the rod at $x = 0$. The problem is to determine the flow of neutrons to the left and to the right of any point x in the rod. In particular, the flow to the left at $x = 0$, which is called the *reflected flow*, and the flow to the right at $x = a$, which is called the *transmitted flow*, are important quantities.

To construct a model we need to know when the interaction between a neutron and an atom in the rod takes place. To this end, we make the following assumption:

> **Assumption (Q)**:
>
> *When a neutron passes through a small (positive) distance δx to the left or to the right of any point x in the rod, then the probability that interaction will occur is $p\delta x$ and the probability that no interaction will occur is $1 - p\delta x$.*

Let us define the following two variables:

$$y(x) = \text{ neutron flow to the left at any point } x \text{ in the rod;}$$
$$z(x) = \text{ neutron flow to the right at any point } x \text{ in the rod.}$$

We shall assume that the functions $y(x)$ and $z(x)$ are differentiable with respect to x on the closed interval $[0, a]$ with continuous first derivatives.

Consider an arbitrary small element of the rod $[x, x+\delta x]$ at an arbitrary point x in the rod. To construct a mathematical model, we note that $y(x)$ is the sum of the contribution of three components:

(1) the neutrons moving to the left at point $x + \delta x$ without any interaction over the closed interval $[x, x + \delta x]$;

(2) the neutrons moving to the left at the point $x + \delta x$ which had interacted over the closed interval $[x, x + \delta x]$;

(3) the neutrons moving to the right at the point x which had interacted over the closed interval $[x, x + \delta x]$.

Using Assumption (Q) we can express this mathematically as

$$y(x) = (1 - p\delta x)y(x + \delta x) + 2p\delta x y(x + \delta x) + p\delta x z(x)$$

and hence

$$y(x + \delta x) - y(x) = -py(x + \delta x)\delta x - pz(x)\delta x \tag{5.4.1}$$

In exactly the same way, when we consider the flow to the right at $x+\delta x$, we find

$$z(x + \delta x) = (1 - p\delta x)z(x) + 2p\delta x z(x) + p\delta x y(x + \delta x)$$

and hence

$$z(x + \delta x) - z(x) = py(x + \delta x)\delta x + pz(x)\delta x \tag{5.4.2}$$

Divide (5.4.1) and (5.4.2) by δx and let δx tend to zero to obtain

$$\left.\begin{aligned} \frac{dy}{dx} &= -py(x) - pz(x) \\ \frac{dz}{dx} &= py(x) + pz(x) \end{aligned}\right\} \tag{5.4.3}$$

For the boundary values, let us consider the case when there is a constant flow α to the right entering the rod at $x = 0$, but no flow to the left entering the rod at $x = a$:

$$\left.\begin{aligned} y(a) &= 0 \\ z(0) &= \alpha \end{aligned}\right\} \tag{5.4.4}$$

The important unknown quantities are the reflected flow $y(0)$ and the transmitted flow $z(a)$. Let us first determine the solution of (5.4.3), (5.4.4) with the aid of Laplace transforms. We shall write $Y(s) = L[y(x)]$ and $Z(s) = L[z(x)]$. Then by Theorem 4.2.2 the equations in (5.4.3) are transformed to

$$\left.\begin{array}{rcl} sY(s) - y(0) & = & -pY(s) - pZ(s) \\ sZ(s) - \alpha & = & pY(s) + pZ(s) \end{array}\right\}$$

which is a system of simultaneous linear equations with the solution

$$\begin{array}{rcl} Y(s) & = & \frac{1}{s}y(0) - p[y(0) + \alpha]\frac{1}{s^2} \\ Z(s) & = & \frac{1}{s}\alpha - p[y(0) + \alpha]\frac{1}{s^2} \end{array}$$

Use Table 4.4.1, number 1, to find the inverse transform:

$$\left.\begin{array}{rcl} y(x) & = & y(0) - p[y(0) + \alpha]x \\ z(x) & = & \alpha + p[y(0) + \alpha]x \end{array}\right\} \tag{5.4.5}$$

Note that $y(0)$ is unknown - by (5.4.4) we only know the value of $y(a)$. Substituting $x = a$ in the first equation of (5.4.5), we find that

$$0 = y(a) = y(0)[1 - pa] - pa\alpha \tag{5.4.6}$$

Now, if $pa = 1$, then (5.4.6) states that $\alpha = 0$. In this case we have a contradiction if the given flow α is nonzero. (The case $\alpha = 0$ is very uninteresting since nothing happens - there is no flow of neutrons at all!) To avoid this dilemma, let us make the following assumption:

Assumption (R)

The length a of the rod is less than $\frac{1}{p}$.

Under the Assumption (R) it follows that $1 - pa > 0$, and hence, we can solve for $y(0)$ in (5.4.6) to obtain

$$y(0) = \frac{pa\alpha}{1 - pa}$$

Substitute this in (5.4.5):

$$\left.\begin{array}{rcl} y(x) & = & \frac{p\alpha}{1-pa}(a - x) \\ z(x) & = & \frac{\alpha}{1-pa}(1 - pa + px) \end{array}\right\} \tag{5.4.7}$$

We have thus found a smooth solution (5.4.7) to the boundary value problem (5.4.3), (5.4.4) under the Assumptions (Q) and (R).

We must still settle the question of uniqueness. Unfortunately this does not follow by Theorem 5.1.1 because the two boundary values in (5.4.4) are given at the two endpoints and not both at the same endpoint as in (5.1.5) of Theorem 5.1.1. This is very important in the proof of Theorem 5.1.1 - in fact (5.10.3) will not hold for two point boundary values like (5.4.4). Hence, we shall have to devise a separate proof for the uniqueness of (5.4.3), (5.4.4).

Suppose that $\bar{y}(x)$ and $\bar{z}(x)$ are a solution of the boundary value problem (5.4.3), (5.4.4), and let $y(x)$ and $z(x)$ be the functions given in (5.4.7).

Definition 5.4.1

The Wronskian $W(x)$ of the two pairs of functions $\{y(x), z(x)\}$ and $\{\bar{y}(x), \bar{z}(x)\}$ is defined as

$$W(x) = y(x)\bar{z}(x) - \bar{y}(x)z(x) \tag{5.4.8}$$

We now have

$$\begin{aligned} \frac{dW}{dx} &= \frac{dy}{dx}\bar{z} + y\frac{d\bar{z}}{dx} - \frac{d\bar{y}}{dx}z - \bar{y}\frac{dz}{dx} \\ &= -pz\bar{z} - py\bar{z} + yp\bar{y} + yp\bar{z} + p\bar{z}z + p\bar{y}z - \bar{y}py - \bar{y}pz \\ &= 0 \end{aligned}$$

By definition (see §5.1) the solution components $\bar{y}(x)$ and $\bar{z}(x)$ are continuous, and hence, $W(x)$ is also continuous on the closed interval $[0, a]$. Hence

$$W(x) = \text{constant} \qquad \text{on } [0, a].$$

By (5.4.4) it follows that

$$\begin{aligned} W(a) &= 0.\bar{z}(a) - 0.z(a) \\ &= 0 \end{aligned}$$

and thus also

$$W(x) = 0 \qquad \text{on } [0, a].$$

In particular at $x = 0$ we have

$$0 = W(0) = y(0)\alpha - \bar{y}(0)\alpha$$

and since $\alpha \neq 0$, it follows that

$$\bar{y}(0) = y(0) = \frac{pa\alpha}{1 - pa}$$

by (5.4.7). Now both solution pairs $\{y(x), z(x)\}$ and $\{\bar{y}(x), \bar{z}(x)\}$ satisfy (5.4.3) and the boundary values

$$\begin{aligned} \bar{y}(0) &= y(0) = \frac{pa\alpha}{1 - pa} \\ \bar{z}(0) &= z(0) = \alpha. \end{aligned}$$

Then by Theorem 5.2.1 we know that

$$\bar{y}(x) = y(x) \text{ and } \bar{z}(x) = z(x) \qquad \text{on } [0, a],$$

and hence the solution (5.4.7) is unique.

Finally, we can show the reflected and transmitted flows:

$$\begin{aligned} y(0) &= \frac{pa\alpha}{1 - pa} \\ z(a) &= \frac{\alpha}{1 - pa} \end{aligned}$$

Note that when the length a of the rod tends to $\frac{1}{p}$, then $1 - pa$ tends to zero, and hence, $y(0)$ and $z(a)$ become unbounded. For this reason the length $\frac{1}{p}$ is called the *critical length*.

- Do: Exercises 8, 9 in §5.8.
- Projects D, E in §5.9.

5.5 Electrical Networks

The analysis of electrical networks is a well-known example of a mathematical model with differential equations. Once the basic theory is understood, it is relatively simple to construct a mathematical model of a given network. To illustrate this, we shall look at a few typical examples in this section.

Consider the two elementary networks in Figures 5.5.1 and 5.5.2 where the *voltage source* (usually a battery or generator) is shown as a circle next to the letter E. When the switch S is closed, the voltage source causes a current i to flow in the network. Both these networks are examples of a *loop,* which just means a closed path (in other words, the initial point and the endpoint coincide) in a network.

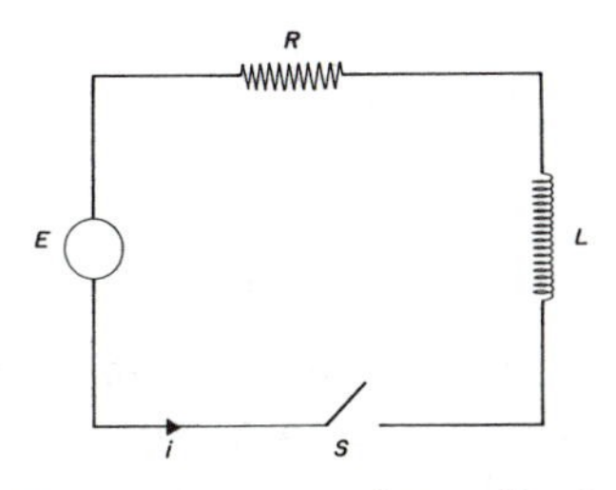

Figure 5.5.1: **Elementary resistor/inductor network**

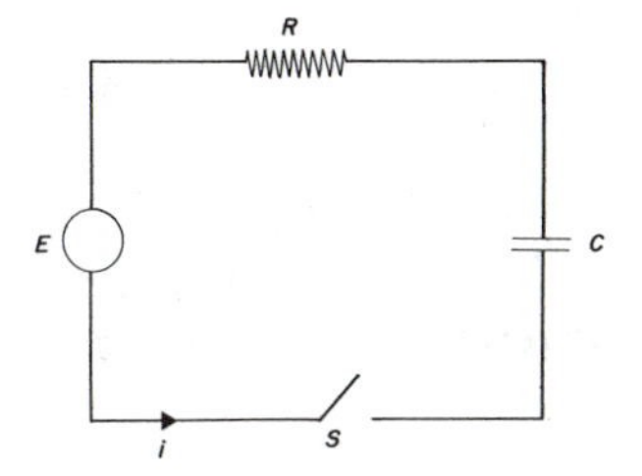

Figure 5.5.2: **Elementary resistor/capacitor network**

A *resistor* (for example, an electric bulb or a heater) opposes the flow of the current, and is shown in both networks as a sawtooth diagram next to the letter R. The change in voltage produced by the resistor is directly proportional to the current, and this proportionality constant is called the *resistance* R of the resistor.

When current passes through an *inductor* a magnetic field is produced which opposes any change in the current. An inductor is shown as the coil-like diagram next to the letter L in Figure 5.5.1. The change in voltage produced by the inductor is directly proportional to the rate of change of the current, and this proportionality constant is called the *inductance* L of the inductor.

A *capacitor* basically consists of two metal plates, separated by material which is a poor conductor of electricity. It is shown as two parallel lines next to the letter C in Figure 5.5.2. When current flows in the network, the capacitor stores electrons on its plates, with the effect of reversing the direction of the current (much like a closed floodgate in a canal). When the charge of electrons on the plates becomes too large for the capacity of the capacitor, then the electrons jump through the isolating material by means of a spark. The change in voltage produced by the capacitor is directly proportional to the charge of the electrons on the plates, and the reciprocal of this proportionality constant is called the *capacitance* C of the capacitor.

Let $q(t)$ denote the charge on the plates of the capacitor at any time t. By definition the current $i(t)$ is given by

$$i(t) = \frac{dq}{dt} \tag{5.5.1}$$

provided that q is measured in coulomb, i in ampere, and the time t in seconds. Then, as was already stated above, we have the following voltage drops for the elements in the networks of Figures 5.5.1 and 5.5.2:

The change in voltage

- *over a resistor of R ohm is Ri volt;*
- *over an inductor of L henry is $L\frac{di}{dt}$ volt;*
- *over a capacitor of C farad is $\frac{1}{C}q$ volt.*

The other important building blocks for a mathematical model of a network are the two laws of Kirchhoff:

(1) **Kirchhoff's current law**: The algebraic sum of the currents at any junction point in a network is zero.

(2) **Kirchhoff's voltage law**: The algebraic sum of the voltages around any loop in a network is zero.

To apply these laws correctly, the following conventions must be observed:

Conventions

(a) If a current is denoted by i in one direction, then the same current in the opposite direction is denoted by $-i$. In particular, currents entering any junction point in a network are taken as positive and currents leaving the junction are taken as negative.

(b) When a loop is traversed in the same direction as the current, then the voltages of all the elements except the source are taken as positive. Conversely, the voltages are taken as negative when traversing in the opposite direction as that of the current.

(c) The voltage of the source is always taken as negative when the loop is traversed in the same direction as the current, and conversely.

Let us apply these laws and conventions to the networks shown in Figures 5.5.1 and 5.5.2. By Kirchhoff's voltage law we have

$$L\frac{di}{dt} + Ri - E = 0$$

If the switch S was closed at time $t = 0$, then the following initial value problem results for the network of Figure 5.5.1:

$$\left.\begin{aligned} \frac{di}{dt} + \frac{R}{L}i &= \frac{E}{L} \qquad (0 < t < \infty) \\ i(0) &= 0 \end{aligned}\right\} \tag{5.5.2}$$

Similarly, the network in Figure 5.5.2 can be modelled by the following initial value problem:

$$\left.\begin{aligned} R\frac{dq}{dt} + \frac{1}{C}q &= E \qquad (0 < t < \infty) \\ q(0) &= \alpha \end{aligned}\right\} \tag{5.5.3}$$

where α is the initial charge on the plates of the capacitor.

Both the differential equations in (5.5.2) and (5.5.3) are linear, and hence, their respective solutions can easily be found with the aid of an integrating factor (see (2.1.13) in §2.1).

- Read: [50] page 172.
- Do: Exercises 10, 11, 12 in §5.8.

Let us now consider a more complicated network, as shown in Figure 5.5.3. There are three possible loops in this network, which are labelled $JHGBNDAKJ$, $JHGBMFAKJ$, and $NBMFADN$ in Figure 5.5.3. Each loop will be represented by a differential equation in our mathematical model in the same way that the loops in Figures 5.5.1 and 5.5.2 were represented by the differential equations (5.5.2) and (5.5.3), respectively. We shall, however, see that any one of the three differential equations is a linear combination of the other two equations; in other words, there are only two linearly independent equations. Naturally, we normally choose the two loops with the simplest differential equations! The first step is to designate directions for the currents i_1, i_2, and i_3 in the different loops, as shown by arrows in Figure 5.5.3. The specific direction chosen for each current is really immaterial, since the sign of the answer will indicate what the real direction of the current is; but we need these directions to apply Kirchhoff's laws.

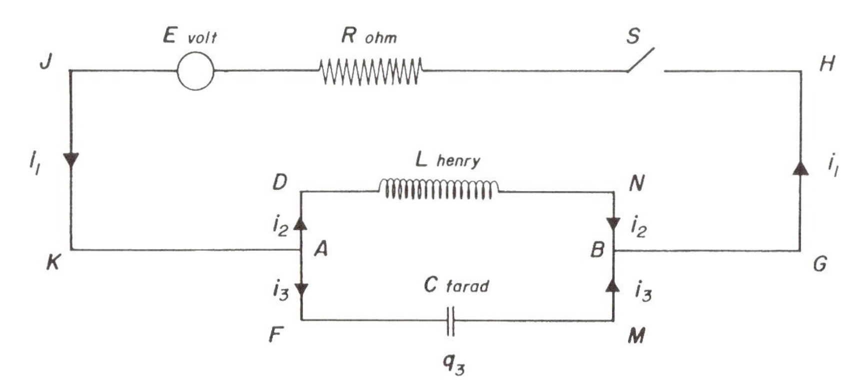

Figure 5.5.3: **Network with more than one loop**

Now use Kirchhoff's current law and the convention (a) at the junction point A or B:

$$i_1 = i_2 + i_3 \tag{5.5.4}$$

Next we use Kirchhoff's voltage law and the conventions (b) and (c) for the loop $JHGBNDAKJ$:

$$E - Ri_1 - L\frac{di_2}{dt} = 0 \tag{5.5.5}$$

Similarly for the loop $JHGBMFAKJ$:

$$E - Ri_1 - \frac{1}{C}q_3 = 0 \tag{5.5.6}$$

with q_3 the charge on the capacitor due to the current i_3. Finally, for the loop $NBMFADN$:

$$-\frac{1}{C}q_3 + L\frac{di_2}{dt} = 0 \tag{5.5.7}$$

Note that (5.5.6) minus (5.5.7) equals (5.5.5) as we predicted. We choose to use (5.5.6) and (5.5.7) in the model. Suppose that q_3 is zero at time

$t = 0$ when the switch is closed. Substitute (5.5.1) and (5.5.4) in (5.5.6) to obtain the following initial value problem:

$$\left.\begin{array}{rclrcl} \dfrac{dq_3}{dt} & = & -\dfrac{1}{RC}q_3 - i_2 + \dfrac{1}{R}E(t) \ , & q_3(0) & = & 0 \\ \dfrac{di_2}{dt} & = & \dfrac{1}{LC}q_3 & i_2(0) & = & 0 \end{array}\right\} \quad (5.5.8)$$

where R, C, and L are positive constants and $E(t)$ is a known continuous function.

Assume that q_3, i_2, and $E(t)$ are of class E with the derivatives of q_3 and i_2 continuous for $t > 0$. Let $Q_3(s) = L[q_3(t)]$ and $I_2 = L[i_2(t)]$. Then by Theorem 4.2.2 the equations in (5.5.8) are transformed to

$$\begin{array}{rcl} sQ_3(s) & = & -\dfrac{1}{RC}Q_3(s) - I_2(s) + \dfrac{1}{R}L[E(t)] \\ sI_2(s) & = & \dfrac{1}{LC}Q_3(s) \end{array}$$

These simultaneous linear equations can easily be solved:

$$\left.\begin{array}{rcl} Q_3 & = & \dfrac{s}{s^2 + \frac{1}{RC}s + \frac{1}{LC}} \ \dfrac{L[E(t)]}{R} \\ I_2(s) & = & \dfrac{1}{s^2 + \frac{1}{RC}s + \frac{1}{LC}} \ \dfrac{L[E(t)]}{LCR} \end{array}\right\} \quad (5.5.9)$$

Let us assume that

$$4R^2C < L \quad (5.5.10)$$

so that the discriminant of the quadratic polynomial in the denominator of (5.5.9) is positive. Then the zeros p and r of this polynomial, namely

$$\begin{array}{rcl} p & = & -\dfrac{1}{2RC} + \dfrac{1}{2}\sqrt{\dfrac{1}{R^2C^2} - \dfrac{4}{LC}} \\ r & = & -\dfrac{1}{2RC} - \dfrac{1}{2}\sqrt{\dfrac{1}{R^2C^2} - \dfrac{4}{LC}} \end{array}$$

are both negative real numbers. Hence by partial fractions

$$Q_3(s) = \frac{1}{p-r}\left(\frac{p}{s-p} - \frac{r}{s-r}\right)\frac{L[E(t)]}{R}$$

Taking inverse Laplace transforms (see Theorem 4.2.4 and Table 4.4.1) we find that

$$\left.\begin{aligned} q_3(t) &= \tfrac{p}{R(p-r)} E(t) * e^{pt} - \tfrac{r}{R(p-r)} E(t) * e^{rt} \\ &= \tfrac{p}{R(p-r)} \int_0^t E(\tau) e^{p(t-\tau)} d\tau - \tfrac{r}{R(p-r)} \int_0^t E(\tau) e^{r(t-\tau)} d\tau \\ &= \tfrac{1}{R(p-r)} \int_0^t E(\tau)(p e^{p(t-\tau)} - r e^{r(t-\tau)}) d\tau \end{aligned}\right\} \tag{5.5.11}$$

Similarly we find

$$\left.\begin{aligned} i_2(t) &= \tfrac{1}{LCR(p-r)} E(t) * e^{pt} - \tfrac{1}{LCR(p-r)} E(t) * e^{rt} \\ &= \tfrac{1}{LCR(p-r)} \int_0^t E(\tau)(e^{p(t-\tau)} - e^{r(t-\tau)}) d\tau \end{aligned}\right\} \tag{5.5.12}$$

We have demonstrated the existence of a solution (5.5.11), (5.5.12) of the initial value problem (5.5.8) when the inequality (5.5.10) holds. The uniqueness follows immediately by Theorem 5.1.1. The other variables of the network in Figure 5.5.3 can now easily be determined. To find $i_3(t)$, differentiate (5.5.11) by the Leibnitz rule (see [57] page 169 or [64] page 353) and use (5.5.1):

$$i_3(t) = \frac{1}{R(p-r)} \int_0^t E(\tau)(p^2 e^{p(t-\tau)} - r^2 e^{r(t-\tau)}) d\tau + \frac{E(t)}{R}$$

Then $i_1(t)$ follows immediately by (5.5.4) and (5.5.12):

$$i_1(t) = \frac{1}{LCR(p-r)} \int_0^t E(\tau)((1+LCp^2)e^{p(t-\tau)} - (1+LCr^2)e^{r(t-\tau)}) d\tau + \frac{E(t)}{R}$$

When (5.5.10) is false, the cases $4R^2C = L$ and $4R^2C > L$ are done in exactly the same way, except of course that the inverse Laplace transforms will be different due to the fact that in the equations (5.5.9) $p = r$ in the one case, and p and r are complex numbers in the other case.

- Read: [58] page 238, [62] page 451. If you are familiar with matrix theory, see also [46] page 102 and [21] page 65.

- Do: Exercises 13, 14, 15 in §5.8.

5.6 Marriage

A simple model of population growth was discussed in §2.2. This Malthus model for the population $N(t)$ at time t can be written

$$\frac{dN}{dt} = (b - d)N$$

where b and d denote the birth rate and the mortality rate, respectively. In general, these birth and mortality rates will differ in the population; for example, the birth rate will be close to zero in a home for senior citizens (let us not underestimate them!). In this section we shall look at subsets of the population and their interaction in the process of population growth. Let us subdivide the population into three classes: (i) unmarried males, (ii) unmarried females, and (iii) married persons.

Assumption (S)

(1) The marriage is monogamous.
(2) A person who was married, but is now single due to divorce or the death of the husband/wife, is unmarried.
(3) Only married couples produce children.

Note that married couples produce children who are unmarried, and conversely, unmarried persons decide to get married, so that the growth of each of the three classes is interrelated. To construct the model, let us define the variables

$m(t)$ = number unmarried males in the population at time t;
$f(t)$ = number unmarried females in the population at time t;
$w(t)$ = number married couples in the population at time t.

Now the rate of change of $m(t)$ depends on (i) the mortality rate a of the unmarried male population, (ii) the rate b at which unmarried females get married, (iii) the birth rate c of (unmarried!) males produced by the married couples, and (iv) the mortality rate d of married females. Hence

$$\frac{dm}{dt} = (c + d)w(t) - am(t) - bf(t) \tag{5.6.1}$$

Similarly the rate of change of $f(t)$ depends on (i) the mortality rate $\bar{a}$ of the unmarried female population, (ii) the rate b at which unmarried females get married, (iii) the birth rate $\bar{c}$ of females produced by the married couples, and (iv) the mortality rate $\bar{d}$ of married males. Hence

$$\frac{df}{dt} = (c + \bar{d})w(t) - (\bar{a} + b)f(t) \tag{5.6.2}$$

Finally, the rate of change of $w(t)$ depends on (i) the rate b at which unmarried females get married, (ii) the mortality rate d of married females, and (iii) the mortality rate $\bar{d}$ of married males. Hence

$$\frac{dw}{dt} = bf(t) - (d + \bar{d})w(t) \tag{5.6.3}$$

With appropriate initial values $m(0), f(0)$, and $w(0)$ given, the equations (5.6.1), (5.6.2), and (5.6.3) can be solved by Laplace transforms. The uniqueness of the solution is guaranteed by Theorem 5.1.1.

- Read: [31] page 73.
- Do: Exercises 16, 17 in §5.8.

5.7 Residential Segregation

Let us suppose that two distinct cultural groups, the Machos and the Nerds, live within the same city. Due to their cultural differences, one group may irritate the other group so much that the one group may start moving out of the city, which then leads to segregation. This situation was modelled in [27] by M. E. Gurtin in the following way:

Let $m(t)$ and $n(t)$ be the number of Machos and Nerds, respectively, at time t in the city. As in §2.2 (see (2.2.1)), we shall restrict ourselves to growth functions which only depend on the population of the two groups:

$$\begin{aligned} \frac{dm}{dt} &= f(m,n) \\ \frac{dn}{dt} &= g(m,n) \end{aligned}$$

We need the following assumptions:

Assumption (T)

(1) *Both groups can move freely in and out of the city.*
(2) *The functions $m(t)$ and $n(t)$ have continuous first derivatives for $t > 0$.*
(3) *The rate of change of the population of one group is directly proportional to the population of both groups.*
(4) *For both groups, the presence of the other group is uncomfortable.*

The Assumption (T3) means that the functions $f(m,n)$ and $g(m,n)$ are linear functions of m and n. We thus obtain the initial value problem

$$\left.\begin{aligned} \frac{dm}{dt} &= am(t) + bn(t) & m(0) &= \alpha \\ \frac{dn}{dt} &= cm(t) + dn(t) & n(0) &= \beta \end{aligned}\right\} \tag{5.7.1}$$

where a, b, c, d, α, and β are constants. Note that a and d are the net growth rate per population, as was the constant k in §2.2. The constants b and c indicate, respectively, the effect of the presence of the other group. By Assumption (T4) both these constants must be negative:

$$b < 0 \text{ and } c < 0 \tag{5.7.2}$$

Let us assume that both $m(t)$ and $n(t)$ are of class E, and use Laplace transforms to solve the system (5.7.1). As before, we shall use capital letters $M(s) = L[m(t)]$ and $N(s) = L[n(t)]$ to denote the Laplace transforms. By Theorem 4.2.2

$$\begin{aligned} (s-a)M - bN &= \alpha \\ -cM + (s-d)N &= \beta \end{aligned}$$

It is a simple matter to solve now for M and N:

$$\left.\begin{aligned} M(s) &= \frac{\alpha s - \alpha d + \beta b}{(s-a)(s-d) - bc} \\ N(s) &= \frac{\beta s - \beta a + \alpha c}{(s-a)(s-d) - bc} \end{aligned}\right\} \tag{5.7.3}$$

Before we can transform back to the t-domain, we must first investigate the zeros of the denominator.

$$s^2 - (a+d)s + ad - bc = 0$$

$$\begin{aligned} s &= \tfrac{1}{2}(a+d) \pm \tfrac{1}{2}\sqrt{(a+d)^2 - 4(ad-bc)} \\ &= \gamma \pm \omega \end{aligned}$$

where we have introduced for convenience the symbols

$$\left.\begin{aligned} \gamma &= \tfrac{1}{2}(a+d) \\ \delta &= \tfrac{1}{2}(a-d) \\ \omega &= \sqrt{\delta^2 + bc} \end{aligned}\right\} \qquad (5.7.4)$$

Note that (5.7.2) guarantees that ω is a real number. Hence we can rewrite the first equation in (5.7.3) as follows:

$$M(s) = \frac{\alpha s - \alpha d + \beta b}{(s-\gamma-\omega)(s-\gamma+\omega)}$$

To find the inverse transform, we first use partial fractions:

$$M(s) = \frac{\alpha(\delta+\omega)+\beta b}{2\omega}\frac{1}{s-\gamma-\omega} - \frac{\alpha(\delta-\omega)+\beta b}{2\omega}\frac{1}{s-\gamma+\omega}$$

Similarly we find that

$$N(s) = \frac{\beta(\omega-\delta)+\alpha c}{2\omega}\frac{1}{s-\gamma-\omega} - \frac{\beta(\omega+\delta)+\alpha c}{2\omega}\frac{1}{s-\gamma+\omega}$$

The inverse transform follows immediately by Table 4.4.1, number 2:

$$\left.\begin{aligned} m(t) &= \tfrac{1}{2\omega}e^{\gamma t}[\{\alpha(\delta+\omega)+\beta b\}e^{\omega t} - \{\alpha(\delta-\omega)+\beta b\}e^{-\omega t}] \\ n(t) &= \tfrac{1}{2\omega}e^{\gamma t}[\{\beta(\omega-\delta)+\alpha c\}e^{\omega t} + \{\beta(\delta+\omega)-\alpha c\}e^{-\omega t}] \end{aligned}\right\} \qquad (5.7.5)$$

Hence, there exists a solution of the initial value problem (5.7.1). Furthermore, by Theorem 5.1.1 we also know that (5.7.5) is the only solution of (5.7.1). Segregation will occur when either $m(t)$ or $n(t)$ is zero in a finite time τ. Now $m(\tau) = 0$ implies that

$$\{\alpha(\delta+\omega)+\beta b\}e^{\omega\tau} = \{\alpha(\delta-\omega)+\beta b\}e^{-\omega\tau}$$

from which τ can be obtained:

$$e^{2\omega\tau} = \frac{1 + \frac{\delta-\omega}{b}\frac{\alpha}{\beta}}{1 + \frac{\delta+\omega}{b}\frac{\alpha}{\beta}} \qquad (5.7.6)$$

Note that $\omega > |\delta|$, and hence $\delta - \omega$ is negative irrespective of the sign of δ. Since b is negative by (5.7.2), and α and β are positive, the expression above the line is positive. For τ to be a real number in (5.7.6), the expression below the line must also be positive, which will be the case if

$$\frac{\alpha}{\beta} < \frac{-b}{\delta + \omega} \tag{5.7.7}$$

In exactly the same way (see Exercise 18 in §5.8) we can show that $n(t)$ will be zero in a finite time if

$$\frac{\alpha}{\beta} > -\frac{\omega - \delta}{c} = \frac{-b}{\delta + \omega} \tag{5.7.8}$$

where the equality follows by (5.7.4). By (5.7.7) and (5.7.8) we have now established that segregation will occur within a finite time whenever

$$\frac{\alpha}{\beta} \neq \frac{-b}{\delta + \omega}$$

When the equality sign holds, we have

$$m(t) = \alpha e^{(\gamma - \omega)t}$$
$$n(t) = \beta e^{(\gamma - \omega)t}$$

so that neither population can be zero in finite time. M. E. Gurtin calls the critical value $A = \frac{-b}{\delta + \omega}$ the *tipping ratio,* because if $\frac{\alpha}{\beta} < A$ then the inhabitants of the city will only be Nerds within a finite time τ, and conversely, if $\frac{\alpha}{\beta} > A$ then the inhabitants of the city will only be Machos after some other finite time (see Exercise 18 in §5.8). When both population groups have the same natural increase $a = d$, then $\delta = 0$ and the tipping ratio is independent of a:

$$A = \sqrt{\frac{b}{c}} \tag{5.7.9}$$

This model can be applied to the urban population of the United States of America. In a study by Duncan and Duncan [19], no instance was found between 1940 and 1950 of a mixed neighborhood with a 25% to 75% white population in which succession from white to black was arrested or reversed.

In another study [25] Grodzins estimates that whites begin to evacuate a neighborhood when it becomes 20% black. Let us assume that P people live in a neighborhood of which $0.2P$ are black and $0.8P$ are white. Then the tipping ratio is

$$A = \sqrt{\frac{b}{c}} = \frac{0.2P}{0.8P} = \frac{1}{4}$$

by (5.7.9). This means that $c = 16b$, so that the rate at which blacks repel whites is sixteen times the rate at which whites repel blacks.

- Do: Exercises 18, 19 in §5.8.

5.8 Exercises

(1) Show that the functions $u_1 = k\cos t, \quad u_2 = k\sin t$ with k an arbitrary real number, satisfy the boundary value problem

$$\begin{aligned} \frac{du_1}{dt} &= -u_2, & u_1(\tfrac{\pi}{2}) &= 0 \\ \frac{du_2}{dt} &= u_1, & u_2(0) &= 0 \end{aligned}$$

on the interval $[0, \frac{\pi}{2}]$. Why is Theorem 5.1.1 not valid in this case?

(2) State and prove a theorem on the continuous dependence on the boundary values $\boldsymbol{\alpha}$ of the solution $\boldsymbol{u}$ of the system (5.1.2), (5.1.3). (Hint: Read first the proof of Theorem 5.1.1 in §5.10, and then use the Gronwall lemma, as in the proof of Theorem 2.4.1.)

(3) Use the mathematical model of §5.2 to find equations from which the range X of the projectile and the total time T for the flight can be calculated. Show that

$$(mg + kV\sin\theta)X = mgTV\cos\theta$$

(4) Construct a mathematical model for the trajectory of a projectile in vacuum. As in §5.2, obtain expressions for the velocity and displacement at any time instant during the flight, as well as the trajectory and the height. Compare your answers with (5.2.4), (5.2.5), (5.2.6), and (5.2.7). (Hint: use the Maclaurin series to verify that your results are, in fact, a special case of these equations.)

(5) Write down the equation for the range of the projectile in §5.2. Let θ be the angle for which the range will be a maximum. For a given k, m, and V find an equation which contains the angle θ. Use a suitable algorithmto compute this angle θ in degrees to two decimal places when $V = 1250\ m/s,\ g = 9.81\ m/s^2$, and

(i) $\frac{k}{m} = 0.05$ (ii) $\frac{k}{m} = 0.1$ (iii) $\frac{k}{m} = 0.001$

(6) Solve the differential equation (5.3.6) with the initial value $z(\alpha) = \beta$ directly to obtain (5.3.5).

(7) Consider the model in §5.3 with the difference that the more William loves Zelda, the more Zelda loves William, and conversely. Suppose that Zelda is neutral when William falls in love with her. Solve this model and draw a graph of $w(t)$ and $z(t)$. Will Zelda's love for William ever exceed William's love for Zelda?

(8) Consider the model in §5.4 for the case when the original neutron is replaced by two others, one moving to the left and the other to the right (see [7] page 14). Solve this model and show that, in this case, there is again a certain critical length.

(9) In a dialyser (or kidney machine) blood is purified by a liquid called the dialysate. In [13] a simple model is constructed for this process. Suppose blood of the patient flows at a rate of Q milliliter per minute past a membrane. At the same time the dialysate flows at a rate of q milliliters per minute in the opposite direction on the other side of the membrane. Let x denote the distance along the membrane, with $x = 0$ being the point where the blood enters and $x = a$ being the point where the blood leaves the membrane. Let $U(x)$ denote the concentration of impurities in the blood and $u(x)$ the concentration of impurities in the dialysate. Since $U(x) > u(x)$, the impurities will move through the membrane from the blood to the dialysate. The following model results (see [13] for details):

$$\begin{aligned} \frac{dU}{dx} &= \frac{k(u-U)}{Q} & \qquad U(0) &= \alpha \\ \frac{du}{dx} &= \frac{k(u-U)}{q} & \qquad u(a) &= 0 \end{aligned}$$

with k a positive constant.

(a) Determine the solution of this mathematical model.

(b) The efficiency of the dialyser is measured by the quantity V called the "clearance":

$$V = \frac{Q}{U(0)}[U(0) - U(a)]$$

Investigate the maximal value for V as (i) a function of a (with fixed Q, q, and k) and (ii) a function of q (with fixed Q, k, and a). How does your answer affect the optimal design of a dialyser?

(10) A loop consists of a switch, a voltage source of 100 volt, a resistor of 10 ohm, and an inductor of 2 henry. If the switch is closed at time $t = 0$, determine the current $i(t)$ in ampère at any time t seconds afterwards. Sketch a rough graph of the function $i(t)$.

(11) A loop consists of a switch, a voltage source of 50 volt, a resistor of 25 ohm, and a capacitor of 0.005 farad. If the switch is closed at time $t = 0$ when the charge in the capacitor is zero, determine the charge and current at any time t seconds afterwards.

(12) A resistor of R ohm varies in time such that

$$R = 1 + 0.02t \qquad (0 \leq t \leq 1000)$$

where t is measured in seconds. This resistor is part of a loop which also contains a capacitor of 0.1 farad and a voltage source of 100 volt. When the switch is closed at time $t = 0$, the capacitor has a charge of 5 coulomb. Determine the charge and current at any time t.

(13) When the switch S is closed at time $t = 0$ in the network shown in Figure 5.8.1, the charge in the capacitor is 1 coulomb and the current through the inductor is zero. Determine the charge on the capacitor and the currents in the different loops of the network at any time t. In particular, calculate each of these currents 5 seconds after the switch was closed, and compare your answer with the case when $t \to \infty$.

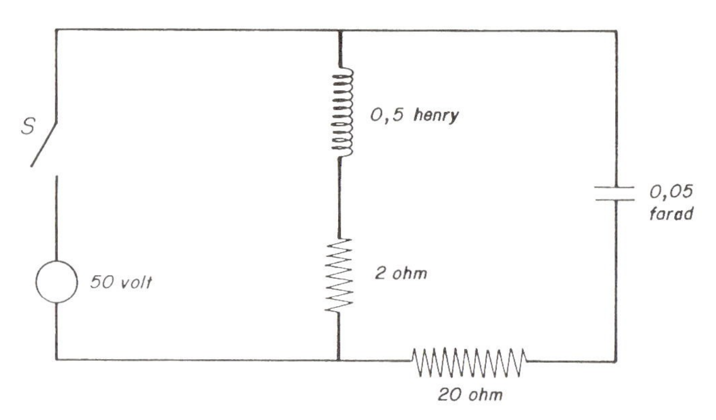

Figure 5.8.1: **Network of Exercise 13**

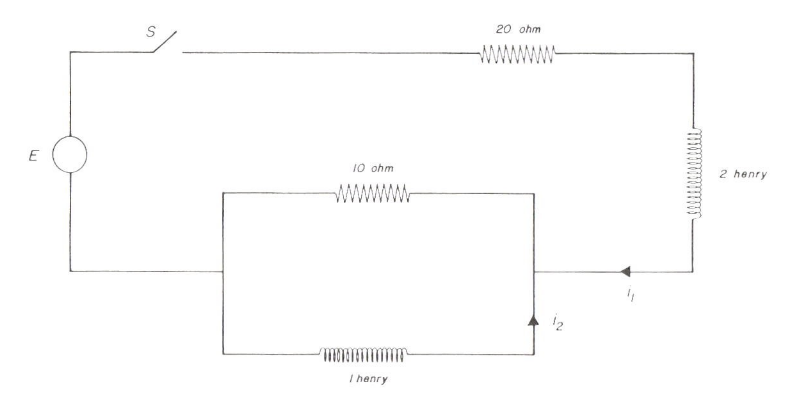

Figure 5.8.2: **Network of Exercise 14**

(14) In the network shown in Figure 5.8.2, the voltage source is $E = 110\cos 10t$ volt. Determine the currents $i_1(t)$ and $i_2(t)$ when the switch is closed at time $t = 0$. What happens when $t \to \infty$?

(15) Solve the model (5.5.8) in §5.5 when $R = 10$ ohm, $C = 10^{-3}$ farad, $E = 100\sin 50t$ volt, and

(a) $L = 0.625$ henry; (b) $L = 0.2$ henry; (c) $L = 0.4$ henry.

(16) In the model (5.6.1) - (5.6.3), assume that $m(0) = f(0) = w(0) = \frac{\alpha}{3}$, where α is the initial population. Let $a = \bar{a} = d = \bar{d} = 9$ per thousand. Solve each of the following two cases:

(i) $c = \bar{c} = 21.6$ per thousand and $b = 10$ per thousand;
(ii) $c = \bar{c} = 59.4$ per thousand and $b = 30$ per thousand.

In each case calculate (a) the population after 10 years and (b) the percentage of each of the classes of the total population at that time.

(17) Construct a mathematical model for the spread of infection of the common cold in an isolated community. Consider three classes in the community at any time t, namely the infectious people, the recuperating people, and the rest who are not ill at that time. Assume that the recuperating people are immune to the illness for a fixed period of time due to the medicine. Can you solve the model by Laplace transforms?

(18) In the model of §5.7 check that (5.7.8) holds. Show also that the segregation time when $n(t) = 0$ is given by

$$\frac{1}{2\omega}\ln(\beta(\delta+\omega)-\alpha c) - \frac{1}{2\omega}\ln(\beta(\delta-\omega)-\alpha c).$$

(19) Investigate the case (5.7.1) where $b = 0$ and $c < 0$ to determine how many Nerds and Machos will be living in the city eventually.

5.9 Projects

Project A^*: Some Ballistics of "Big Bertha"

In World War 1 the extremely big siege guns had a maximal range of about 30 kilometers. Imagine the consternation of the French in 1918 when the Germans began to shell Paris from a distance of 120 kilometers! The secret of "Big Bertha" was closely guarded by the

German military authorities at the time, but today we know that the shells were fired at an initial velocity of about 2000 meters per second at an elevation of 55 degrees, so that the shell travelled high enough for air resistance to be reduced to a minimum. Construct a simplified mathematical model to check the work of the German mathematicians. Use the Assumptions (L) and (N) in §5.2 to obtain (5.2.1) with $k = 0.05$. Let us model the stratosphere by assuming that $k = 0$ at altitudes of more than 10 kilometers (in other words at an altitude of 10 kilometers the density of air changes abruptly to zero).

(a) Calculate the time for the shell to reach the stratosphere, accurate to three decimal places.

(b) Calculate the velocity of the shell when it enters the stratosphere.

(c) Use the velocity in (b) as the initial velocity and the model of the trajectory in vacuum (see Exercise 4 in §5.8), to calculate the maximum altitude of the shell.

(d) Calculate the speed at which the shell enters the statosphere again.

(e) Calculate the time in which the shell then falls (from an altitude of 10 kilometers) to the earth, accurate to three decimal places. (Hint: use the same model as in §5.2 with negative altitudes.)

(f) Calculate the range of "Big Bertha" according to this mathematical model.

(g) Discuss the validity of this mathematical model.

Project B^*: A Square Look at Air Resistance

(a) Consider the mathematical model of §5.1 with Assumption (L) replaced by

> **Assumption (L').**
> *The air resistance is directly proportional to the square of the speed of the projectile at any time during the flight, but in the opposite direction.*

Let the initial velocity be $V = 985$ meters per second at an inclination of $\theta = 50°$, and the constant of proportionality in Assumption (L') be 0.0015.

(b) Write down the equations for the components x' and y' of the velocity at any time t.

(c) Use the Euler algorithm of §3.3 to find x' and y' numerically every second during the flight until the maximum altitude is reached. Plot the values of $x'(t)$ and $y'(t)$.

(d) Find the values of x and y every second during the flight by numerical integration (use the trapezoidal rule - see §3.9) and plot the values of $x(t)$ and $y(t)$.

(e) Use the values of x and y to plot the trajectory of the projectile.

(f) Calculate the speed v and the angle u with the horizontal every second during the flight.

(g) Transform the equations for the components x' and y' of the velocity by

$$\begin{aligned} x' &= v \cos u \\ y' &= v \sin u \end{aligned}$$

Use

$$\frac{dv}{du} = \frac{v'}{u'}$$

to find a first order differential equation in v and u, and rewrite the equation to obtain $\frac{d}{du}(v \cos u)^{-2}$ on the left hand side. Solve this differential equation. Use the solution to check the values of v and u found in (f) above.

Project C^*: Steel Production

In [7] page 21, Richard Bellman suggests the following simplified model for the production of steel: Suppose that a factory has a sufficient stock of raw materials necessary for the production of steel. Denote the maximum capacity (to produce steel) of the factory at time t by $y(t)$ mass per time unit (for example, tons per day). Denote the total production of steel in the interval $[0, t]$ by $x(t)$ and call it the *stockpile* of the factory. If the production capacity is constant, then the stockpile increases by ayT in the interval $[t, t + T]$ where a is a constant between zero and 1. When $a = 1$ the factory produces at its full capacity, and when $a = 0$ there is no production. Obviously, if the production capacity is increased, the stockpile will grow faster. But to increase the capacity, the factory must be enlarged, which, in its turn, requires steel. Steel for this purpose is taken from the stockpile. Let a fixed fraction b of the stockpile be used per time unit at any time t to increase the production capacity. Make the following assumption about the relation between the amount of steel used and the resulting increase in capacity:

> **Assumption (M)**
>
> *The increase in the production capacity is directly proportional to the amount of steel used to enlarge the factory.*

(1) If the initial stockpile is α and the initial production capacity is β, show that

$$\left.\begin{array}{rclrcl} \dfrac{dx}{dt} & = & -bx + ay, & x(0) & = & \alpha \\ \dfrac{dy}{dt} & = & bcx, & y(0) & = & \beta \end{array}\right\} \qquad (5.9.1)$$

where c is a constant.

(2) Show that the unique solution of this initial value problem is

$$\left.\begin{array}{rcl} x(t) & = & \dfrac{p\alpha + a\beta}{p-q}e^{pt} - \dfrac{q\alpha + a\beta}{p-q}e^{qt} \\ y(t) & = & \dfrac{p\beta + b\beta + bc\alpha}{p-q}e^{pt} - \dfrac{q\beta + b\beta + bc\alpha}{p-q}e^{qt} \end{array}\right\} \qquad (5.9.2)$$

where p and q are the zeros of the quadratic function

$$s^2 + bs - abc.$$

(3) The important question is what b should be to maximize the stockpile x over a given period of T years. (Note the similarity between this problem and the maximal profit problem of §2.6.) Show that

$$\begin{aligned} x(T) &= e^{-\frac{bT}{2}}\Big[\frac{2a\beta - b\alpha}{\sqrt{b^2+4abc}}\sinh\Big(\frac{T}{2}\sqrt{b^2+4abc}\Big) \\ &\quad + \alpha\cosh\Big(\frac{T}{2}\sqrt{b^2+4abc}\Big) \end{aligned} \qquad (5.9.3)$$

Use scaling to introduce new variables:

$$\left.\begin{array}{rcl} z & = & \frac{T}{2}\sqrt{b^2+4abc} \\ w & = & x(T)e^{-acT} \end{array}\right\} \qquad (5.9.4)$$

Consider the case when $\alpha = 0$. Try to find an expression for z which will maximize w. Allocate a few values to acT and then try a numerical procedure to find z.

Project D^*: A Combat Model

Since the pioneering work of F. W. Lanchester (see [38] and [39]), several papers were published on the application of a combat model to actual battle situations. For example, J. H. Engel applied the Lanchester model to the Battle of Iwo Jima (see [20]), and a model of the Battle of the Ardennes (more commonly known as the Battle of the Bulge) was constructed by four students of Harvey Mudd College (see [33]). In this project we study a simple model of a combat situation.

Suppose two opposing forces meet in battle. Let $x(t)$ and $y(t)$ denote the force strength of the two armies at time t. We assume that the two sides engage in open combat, so that each side is wholly exposed to the fire of the other side. Under this assumption the combat loss rate of an army will vary linearly with the size of the opposing force, with the proportionality constant indicating the efficiency of the opposing force. In addition both armies will suffer non-combat losses which we shall assume to be constant for each army during the battle. (In general, these losses are taken to be proportional to the size of the army, but in some cases our assumption is realistic - see, for example, [33].)

(a) Show that this battle can be modelled as

$$\left.\begin{aligned} \frac{dx}{dt} &= -ay + f(t) - c & \qquad x(0) &= \alpha \\ \frac{dy}{dt} &= -bx + g(t) - d & \qquad y(0) &= \beta \end{aligned}\right\}$$

where $f(t)$ and $g(t)$ denote the rate of reinforcements for the two armies during the battle.

(b) Assume that $f(t) = k$ and $g(t) = \ell$ are both constant during the battle. Determine the force strength of each army at any time t during the battle.

(c) If
$\alpha > \frac{\ell - d}{b} > 0$ and $\beta > \frac{k-c}{a} > 0$
determine the conditions under which the y-force will be wiped out, and the time taken for such a decisive win of the x-force. Are the assumptions still realistic in this situation?

(d) If
$a = 0.006, b = 0.008, c = d = 1,000, k = 6,000, \ell = 4,000,$

$\alpha = 90,000, \beta = 200,000$
where c, d, k, and ℓ are measured in men per day, draw graphs of $x(t)$ and $y(t)$ on the same axes for $0 \leq t \leq 50$. Use these graphs to determine the time τ when $x(\tau) = y(\tau)$. Which side is winning after 50 days?

Project E^*: The Battle of the Alamo

In the annals of bravery in combat, the Battle of the Alamo is one of the highlights. The heroic stand against impossible odds to win time for regrouping of the main army is an unforgettable example of unselfish sacrifice. In the battle for the independence of the state of Texas, the president of Mexico, General Antonio Lòpez de Santa Anna, was stopped in his advance into Texas by a small force of 188 men (including the legendary Davy Crockett) in the fort at the mission station at the Alamo near San Antonio. The Mexicans attacked the fort with a force of 3000 men on a very cold Sunday morning, 6 March 1836. In the first phase of the battle the Texans were protected in the fort and the Mexicans were out in the open, with the result that the Mexicans suffered severe losses. It is estimated that at the end of the first phase only 1800 Mexican men and 100 Texan men were still fighting. The second phase started when the Mexicans succeeded in breaching the defenses of the fort. The battle changed to a hand-to-hand fight with directed fire. The Texans were wiped out and the Mexicans lost half of their men. In the meantime, General Sam Houston could regroup his men, and he won a decisive victory over the Mexicans at San Jacinto on 21 April 1836.

Let $x(t)$ and $y(t)$ denote the force strength of the Texans and the Mexicans, respectively, at time t where t is measured in hours. During the first phase we can model the battle as

$$\begin{aligned} \frac{dx}{dt} &= -0.0007xy \\ \frac{dy}{dt} &= -ax \end{aligned}$$

where a is the proportionality constant indicating the efficiency of the Texan force (see Project D above). Note that the first equation is nonlinear, so that Laplace transforms cannot be used to find a solution.

During the second phase, the first equation changes to

$$\frac{dx}{dt} = -by$$

$$\frac{dy}{dt} = -ax$$

where b is the proportionality constant indicating the efficiency of the Mexican force, and the second equation is still valid.

(a) Formulate appropriate assumptions to explain why the equations above are used to model the two phases of the battle.

(b) Eliminate the variable t in the equations for the first phase to obtain a first order differential equation in x and y (look again at §5.3 where the phase plane was discussed). Solve this equation and use the data given to calculate the constant a.

(c) Use the same technique as in (b) to calculate the constant b from the equations for the second phase of the battle.

(d) Use (b) to find the force strength of the Mexicans at any time t during the first phase. Draw a graph of $y(t)$ for the first phase.

(e) Use (c) to find the force strength of the Texans at any time t during the second phase. Draw a graph of $x(t)$ for the second phase. Calculate the time that the second phase had lasted.

5.10 Mathematical Background

In this section we prove the uniqueness of a linear system of first order differential equations. Note the close similarity of this proof with the proof of Theorem 2.4.1 in §2.14 where the Gronwall lemma is also used.

Theorem 5.1.1

Suppose that each of the coefficients $b_{ij}(t)$ for $i, j = 1, 2, \ldots, n$ in (5.1.4) are continuous on the closed interval $[a, T]$. If $\boldsymbol{u}$ and $\boldsymbol{v}$ are solutions of the system (5.1.4), (5.1.5), then $\boldsymbol{u}(t) = \boldsymbol{v}(t)$ for all $t \in [a, T]$.

Proof Let $w_i = u_i - v_i$ for $i = 1, 2, 3, \ldots\ldots, n$. Since both $\boldsymbol{u}(t)$ and $\boldsymbol{v}(t)$ satisfy (5.1.4), it follows by subtracting the system for $\boldsymbol{v}$ from the system for u that $w(t)$ satisfies the system

$$\left.\begin{aligned}\frac{dw_1}{dt} &= b_{11}(t)w_1 + b_{12}(t)w_2 + \ldots + b_{1n}(t)w_n\\ \frac{dw_2}{dt} &= b_{21}(t)w_1 + b_{22}(t)w_2 + \ldots + b_{2n}(t)w_n\\ &\ldots \qquad \ldots \qquad \ldots\\ \frac{dw_n}{dt} &= b_{n1}(t)w_1 + b_{n2}(t)w_2 + \ldots + b_{nn}(t)w_n\end{aligned}\right\} \tag{5.10.1}$$

and the boundary values

$$w_1(a) = w_2(a) = \ldots = w_n(a) = 0 \tag{5.10.2}$$

Note that the linearity of the equations (5.1.4) is essential for this step. Since each of the functions $b_{ij}(t)$ where $i, j = 1, 2, ..., n$ is continuous on the closed interval $[a, T]$, they must each be bounded. Let M denote the maximum of all these bounds:

$$M = \max_{i,j=1,2,\ldots,n} \max_{a \leq t \leq T} |b_{ij}(t)|$$

Integrate each of the equations in (5.10.1) and use (5.10.2) to obtain for $i = 1, 2, \ldots, n$

$$w_i(t) = \int_a^t (b_{i1}w_1 + b_{i2}w_2 + \ldots + b_{in}w_n)d\tau \tag{5.10.3}$$

which is valid for all $t \in [a, T]$. Hence

$$\begin{aligned}|w_i(t)| &\leq \int_a^t (|b_{i1}||w_1| + |b_{i2}||w_2| + \ldots + |b_{in}||w_n|)d\tau\\ &\leq M\int_a^t (|w_1| + |w_2| + \ldots + |w_n|)d\tau\end{aligned}$$

Addition of these n inequalities gives

$$|w_1(t)| + |w_2(t)| \ldots + |w_n(t)| \leq Mn\int_a^t (|w_1| + |w_2| + \ldots + |w_n|)d\tau.$$

Note that the expression on the left is continuous by our definition of a solution (see §5.1). We can now apply Gronwall's lemma (see §2.14) with $c = 0$ and $g(t) = 1$ in (2.14.9) to obtain for $a \leq t \leq T$ the inequality

$$|w_1(t)| + |w_2(t)| \ldots + |w_n(t)| \leq 0.$$

Since the left hand side is always non-negative, the inequality implies that

$$w_i(t) = 0 \qquad (i = 1, 2, \ldots\ldots, n)$$

on the interval $[a, T]$, which proves the theorem.

Note that the same proof can be used if the condition that the functions $b_{ij}(t)$ be continuous on the interval $[a, T]$ is relaxed to the condition that the functions $|b_{ij}(t)|$ are bounded on $[a, T]$ and that the integral in (5.10.3) must exist.

6

Second Order Linear Differential Equations

6.1 Mechanical Vibrations

In many mechanical systems the motion is an oscillation with the position of static equilibrium as the center. The suspension of an automobile or the motion of the flywheel in a watch are two well-known examples. Even a solid structure like a bridge or a concrete weir vibrates due to the action of wind, water, or earth tremors. With appropriate assumptions a mathematical model for such an oscillatory motion can be constructed. Depending on the specific situation, the model can be quite complicated. In this section we shall look at a very simple situation which illustrates the essential features of an oscillatory motion.

Consider a spring of length ℓ. A bob and a damping mechanism (shown in Figure 6.1.1 as a so-called dashpot, which is just a cylinder filled with fluid, and a piston, much like a shock absorber) are attached to one end of the spring. The other end is fastened to a vertical hook, as shown in Figure 6.1.1, in such a way that the whole system hangs motionless. Then the spring will lengthen to a length of $\ell + h$ in this static position, where h is a positive real number. Let the total mass on the lower end of the spring be m. When the bob is displaced from the positio of static equilibrium and then projected at a given initial speed in a vertical direction, we would expect the bob to execute an oscillatory motion.

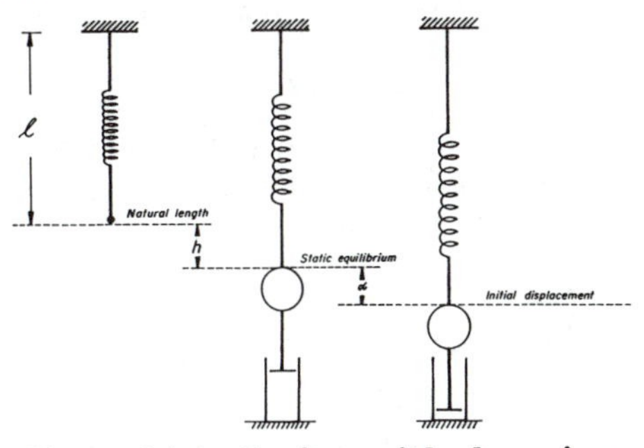

Figure 6.1.1: **Spring with damping**

Let us construct a mathematical model which describes the displacement and velocity of the bob at any time t.

When a force is applied to the lower end of a spring, one feels intuitively that the extension of the spring should bear some relation to the magnitude of the force. This is indeed the case; in fact, it is a linear relationship when the spring is made of steel and the extension h is much smaller than the natural lenght ℓ. This empirical law is due to Robert Hooke (1635 - 1703) and can be formulated as follows:

Hooke's law

The force necessary to extend a spring or an elastic string of natural length ℓ to a length $\ell + h$ is proportional to the extension h, provided that $h \ll \ell$.

The proportionality constant in this law is called the *spring constant* or the *stiffness* and is usually denoted by k. Another constant which is also associated with Hooke's law is called the *modulus* and is usually denoted by λ. The modulus is the product of the spring constant and the natural length of the spring: $\lambda = k\ell$. We shall make the following assumptions:

Assumption (U)

(1) *The mass of the spring is much smaller than the mass of the bob.*

(2) *The amplitude (i. e. the maximal displacement from the center) of the oscillations is so small that it is less than h, and Hooke's law is valid.*

(3) *The magnitude of the resistance to the motion (due to air resistance and the damping mechanism) is directly proportional to the speed at any time t, and always in the opposite direction to the velocity.*

Note that Assumption (U2) implies that the bob is never a free falling body, because the length of the spring is always bigger than the natural length ℓ during the oscillatory motion. By Assumption (U1) we ignore the weight of the spring itself. When the system is hanging in equilibrium, the acceleration is zero. Hence, the resultant force on the bob must also be zero by Newton's second law (see §2.9), or equivalently, the forces above and below the bob must be equal in magnitude:

$$kh = mg \tag{6.1.1}$$

where k is the spring constant and g is the acceleration due to gravity.

Let $x(t)$ denote the displacement of the bob at time t, where the x-axis is chosen vertically downward with the origin at the position of the bob when the system is in static equilibrium. Since the motion is in a straight line, we can describe the direction of a vector merely by a positive or negative sign. We shall denote the derivative of x with respect to t by x'.

There are four different forces acting on the bob at time t:

- the weight mg which is always downward;
- the force $-k(x+h)$ which is always upward (note the minus sign!) by Assumption (U2) since $|x| < h$;
- the resistance (or damping) $-cx'$ which is always in the opposite direction to x, and where c is the proportionality constant in Assumption (U3);
- an external force $f(t)$ directed downwards when its sign is positive, and upwards when its sign is negative.

Hence by Newton's second law (see §2.9) it follows that

$$mx'' = mg - k(x+h) - cx' + f(t)$$

Using (6.1.1) we obtain the following initial value problem:

$$\left.\begin{array}{rcll} mx'' + cx' + kx & = & f(t) & (0 < t < \infty) \\ x(0) & = & \alpha & \\ x'(0) & = & \beta & \end{array}\right\} \tag{6.1.2}$$

where the initial values α and β are given and $m > 0, k > 0, c \geq 0$. We immediately note that the differential equation in (6.1.2) is linear with constant coefficients (compare with (1.2.1) in §1.2).

Before we discuss the existence of a solution of (6.1.2), let us consider the question of uniqueness. Introduce a new variable v to rewrite (6.1.2) as a system of first order differential equations:

$$\left.\begin{aligned} \frac{dx}{dt} &= v \\ \frac{dv}{dt} &= -\frac{c}{m}v - \frac{k}{m}x + \frac{1}{m}f(t) \\ x(0) &= \alpha \\ v(0) &= \beta \end{aligned}\right\} \tag{6.1.3}$$

The uniqueness of a solution of (6.1.3), and hence also of (6.1.2), now immediately follows by Theorem 5.1.1 in §5.1.

To solve the boundary value problem (6.1.2), we shall consider two cases separately: (i) the case when there are no external forces, or the forcing term $f(t) = 0$, called *free motion*, and (ii) the case of *forced motion* when $f(t) \neq 0$. When the right hand side of the differential equation in (6.1.2) is zero, it is also said to be *homogeneous*; similarly in the case of a forced motion, the differential equation is said to be *inhomogeneous*. (This must not be confused with another use of the same word, defined also as homogeneous equations in §2.1.)

(i) Free Motion ($f = 0$)

We must find a solution of the boundary value problem

$$\left.\begin{aligned} mx'' + cx' + kx &= 0 \qquad (0 < t < \infty) \\ x(0) &= \alpha \\ x'(0) &= \beta \end{aligned}\right\} \tag{6.1.4}$$

Since the equation is linear, we can use Laplace transforms (see Exercise 1 in §6.5). Alternatively, we can ask the question whether a function of the form

$$x = e^{\lambda t}$$

with λ a real or complex number, could be a solution of (6.1.4). After substituting this function into the differential equation and using the fact that an exponential function is always positive, we find the *auxiliary equation*

$$m\lambda^2 + c\lambda + k = 0$$

The polynomial in λ is also called the *characteristic polynomial*. The roots of this quadratic equation in λ are

$$p = \frac{-c + \sqrt{c^2 - 4km}}{2m}, \quad q = \frac{-c - \sqrt{c^2 - 4km}}{2m} \tag{6.1.5}$$

We must now distinguish three cases, depending on whether p and q are real, equal, or complex numbers, which in turn depend on the damping factor c:

(a) Overdamping ($c^2 > 4km$)

In this case the discriminant is positive, and thus p and q are negative real numbers. We now have two functions which satisfy the differential equation, and hence by the linearity of the equation, so would a linear combination of the two functions:

$$x(t) = Ae^{pt} + Be^{qt}$$

where A and B are arbitrary real numbers. To satisfy the initial values we must also have

$$\begin{aligned} \alpha &= A + B \\ \beta &= pA + qB. \end{aligned}$$

Solving these two linear equations in A and B we find

$$A = \frac{q\alpha - \beta}{q - p} \text{ and } B = \frac{p\alpha - \beta}{p - q}$$

which then supplies a solution of (6.1.4)

$$x(t) = \frac{(\beta - q\alpha)e^{pt} - (\beta - p\alpha)e^{qt}}{p - q} \tag{6.1.6}$$

which we know is the unique solution of (6.1.4) by Theorem 5.1.1, as we had seen above. The solution (6.1.6) can be written in terms of the constants in (6.1.4) by utilizing the definition of p and q in (6.1.5) as follows:

$$\left.\begin{aligned} x(t) &= [\frac{c\alpha + 2m\beta}{\sqrt{c^2 - 4km}} \sinh(\sqrt{c^2 - 4km} \cdot \frac{t}{2m}) \\ &\quad + \alpha \cosh(\sqrt{c^2 - 4km} \cdot \frac{t}{2m})]e^{-\frac{ct}{2m}} \end{aligned}\right\} \tag{6.1.7}$$

Let us take a closer look at (6.1.6) to see what the motion of the bob will be in this case. First, note that $x(t) \to 0$ as $t \to \infty$, since p and q are both negative. The next question is: how many times would the bob pass through the origin of the x-axis (or the point of static equilibrium)? When the bob passes through the origin at time τ, the displacement must then be zero; hence $x(\tau) = 0$. By (6.1.6)

$$e^{(p-q)\tau} = \frac{\beta - p\alpha}{\beta - q\alpha}$$

Clearly, there can be at most one positive value of τ when the right hand side is bigger than 1. (Note that $p - q$ is a positive real number by (6.1.5).) Then we must have

$$(\beta - p\alpha)(\beta - q\alpha) = \frac{1}{m}(m\beta^2 + c\alpha\beta + k\alpha^2) > 0$$

and

$$|\beta - p\alpha| > |\beta - q\alpha|$$

If these two conditions are not met, the bob would not pass through the origin, but will tend to the origin as $t \to \infty$. If these two conditions do hold, the bob would pass only once through the origin and then tend to the origin as $t \to \infty$. Hence, in the case of overdamping there are no oscillations of the bob.

(b) Critical Damping ($c^2 = 4km$)

In this case $p = q = -\frac{c}{2m}$, and so only one function which satisfies the equation is found. To find another function try te^{pt}, which then does satisfy the differential equation. Hence, the linear combination

$$x(t) = (A + Bt)e^{-\frac{ct}{2m}}$$

will satisfy the differential equation in (6.1.4) for arbitrary values of A and B. By the initial values we find that

$$\alpha = A \text{ and } \beta = -\frac{c}{2m}A + B$$

which then gives the unique solution

$$x(t) = (\alpha + \beta t + \frac{c\alpha t}{2m})e^{-\frac{ct}{2m}} \tag{6.1.8}$$

We could also find (6.1.8) if we let $p \to q$ in (6.1.6). Write (6.1.6) in the form

$$\begin{aligned} x(t) &= \frac{1}{p-q}[(\beta - q\alpha)e^{(p-q)t} - (\beta - p\alpha)]e^{qt} \\ &= \frac{1}{p-q}[(\beta - q\alpha)\{1 + (p-q)t + \frac{(p-q)^2t^2}{2!} + \frac{(p-q)^3t^3}{3!} \\ &\quad +.....\} - (\beta - p\alpha)]e^{qt} \\ &= [\alpha + (\beta - q\alpha)t + \frac{1}{2}(\beta - q\alpha)(p-q)t^2 + ...]e^{qt} \end{aligned}$$

where the Maclaurin series of the exponential function was used (see §2.14). If we let $p \to q$ in this equation, then (6.1.8) follows immediately.

As in the case of overdamping, we also find here that $x(t) \to 0$ as $t \to \infty$. This follows immediately by (6.1.8) and the rule of L'Hôpital (see §6.7). Moreover, if the bob passes through the origin at time $t = \tau$, then $x(\tau) = 0$, which can only occur when

$$\tau = -\frac{2m\alpha}{2m\beta + c\alpha}$$

Hence, if $\beta \neq -\frac{c}{2m}\alpha$, then the bob passes exactly once through the origin before it tends to the origin as $t \to \infty$; otherwise it merely tends to the origin from its initial position. Either way, oscillations do not occur.

(c) Underdamping ($c^2 < 4km$)

In this case p and q are complex conjugate numbers in (6.1.5). Let us write

$$\begin{aligned} p &= -\frac{c}{2m} + iw, \; q = -\frac{c}{2m} - iw \\ \text{where } i &= \sqrt{-1} \text{ and } w \text{ is the positive real number} \\ w &= \frac{\sqrt{4km - c^2}}{2m} \end{aligned} \tag{6.1.9}$$

As before, the following function satisfies the differential equation (6.1.4) for arbitrary values of A and B:

$$\begin{aligned} x(t) &= Ae^{-\frac{ct}{2m}}e^{iwt} + Be^{-\frac{ct}{2m}}e^{-iwt} \\ &= e^{-\frac{ct}{2m}}[A\cos wt + iA\sin wt + B\cos wt - iB\sin wt] \\ &= e^{-\frac{ct}{2m}}[(A + B)\cos wt + i(A - B)\sin wt] \end{aligned}$$

Note that if $A = \frac{1}{2}(d_1 + id_2)$ and $B = \frac{1}{2}(d_1 - id_2)$ are conjugate complex numbers, with d_1 and d_2 arbitrary real numbers, then

$$x(t) = e^{-\frac{ct}{2m}}[d_1 \cos wt + d_2 \sin wt]$$

To satisfy the initial values of (6.1.4), we must have

$$\alpha = d_1 \text{ and } \beta = -\frac{c}{2m}d_1 + wd_2.$$

Solving for d_1 and d_2, we find that a solution is

$$x(t) = e^{-\frac{ct}{2m}}[\alpha \cos wt + \frac{2m\beta + c\alpha}{2mw}\sin wt] \tag{6.1.10}$$

which is the unique solution of (6.1.4) by Theorem 5.1.1, as before. If (6.1.9) is used to substitute for the original constants, the solution looks like this:

$$\begin{aligned} x(t) &= [\frac{c\alpha + 2m\beta}{\sqrt{4km - c^2}} \sin(\sqrt{4km - c^2}\frac{t}{2m}) \\ &\quad + \alpha\cos(\sqrt{4km - c^2} \cdot \frac{t}{2m})]e^{-\frac{ct}{2m}} \end{aligned}$$

Note that the form of this solution is very close to the solution (6.1.7) of overdamping. The solution (6.1.10) can also be rewritten in a very useful form by introducing a *phase angle* as follows: Define the phase angle ϕ by

$$\left.\begin{aligned} \tan\phi &= \frac{2m\beta + c\alpha}{2mw\alpha} \\ &= (4km - c^2)^{-\frac{1}{2}}(c + \frac{2m\beta}{\alpha}) \end{aligned}\right\} \qquad (6.1.11)$$

where $-\pi < \phi \leq \pi$ and the correct quadrant is determined by the sign of α on the horizontal axis and $2m\beta + c\alpha$ on the vertical axis. In terms of the phase angle we can now rewrite (6.1.10) in the more compact form

$$x(t) = De^{-\frac{ct}{2m}}\cos(wt - \phi) \qquad (6.1.12)$$

with D the positive root of

$$\left.\begin{aligned} D^2 &= (\frac{2m\beta + c\alpha}{2mw})^2 + \alpha^2 \\ &= \frac{4m}{4km - c^2}(m\beta^2 + c\beta\alpha + k\alpha^2) \end{aligned}\right\} \qquad (6.1.13)$$

The special case when there is no damping at all, or equivalently when $c = 0$, is known as *simple harmonic motion.* Then (6.1.9) reduces to

$$w_0 = \sqrt{\frac{k}{m}} \qquad (6.1.14)$$

and the differential equation is satisfied by

$$x(t) = d_1\cos w_0 t + d_2\sin w_0 t \qquad (6.1.15)$$

Moreover, (6.1.12) can be written as

$$x(t) = \sqrt{\alpha^2 + \frac{m}{k}\beta^2}\ \cos(wt - \phi)$$

Since the graph of $x(t)$ is a cosine function, the bob oscillates with amplitude

$$\sqrt{\alpha^2 + \frac{m}{k}\beta^2}$$

around the position of static equilibrium $x = 0$. The *period* of the oscillation is defined as the time for a complete oscillation - in this case the time

$$T = \frac{2\pi}{w} = 2\pi\sqrt{\frac{m}{k}}$$

When $c \neq 0$ it follows by (6.1.12) that the bob also oscillates around the position of static equilibrium $x = 0$. However, in this case the "amplitude" $De^{-\frac{ct}{2m}}$ of the oscillation diminishes as the time t increases, as is shown in Figure 6.1.2. As $t \to \infty$ the motion is damped out, and the bob ends in the position of static equilibrium $x = 0$.

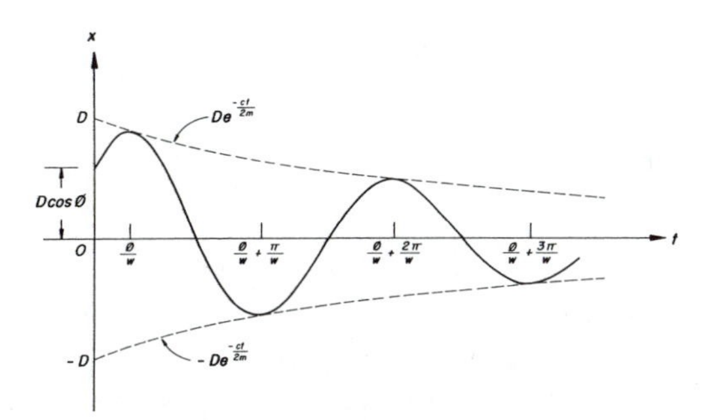

Figure 6.1.2: **Displacement in the case of underdamping**

If we define, as in the case of simple harmonic motion, the "period" as

$$\tau = \frac{2\pi}{w}$$

then

$$\tau = 4m\pi(4km - c^2)^{-\frac{1}{2}} = T(1 - \frac{c^2}{4km})^{-\frac{1}{2}}$$

We immediately note that $\tau > T$. This is not unexpected, since the opposing resistance must retard the motion, so that it takes longer to complete an oscillation.

(ii) Forced Motion ($f \neq 0$)

We now consider the case when an external force acts on the bob, which was modelled as (6.1.2):

$$\begin{aligned} mx'' + cx' + kx &= f(t) \qquad (0 < t < \infty) \\ x(0) &= \alpha \\ x'(0) &= \beta \end{aligned}$$

Let us investigate the case when the external force is periodic:

$$f(t) = E\cos(at + b) \tag{6.1.16}$$

where E and a are positive constants and b is any constant. We shall use the constant w_0 which was defined in (6.1.14) and which had also appeared in the solution of simple harmonic motion. The parameters c and w_0 will now determine the behaviour of the bob. We distinguish three cases:

(a) Undamped Motion: $c = 0, w_0 \neq a$

In this case the damping is so small that it can be ignored, and thus by (6.1.2)

$$\left.\begin{aligned} mx'' + kx &= E\cos(at + b) \qquad (0 < t < \infty) \\ x(0) &= \alpha, \\ x'(0) &= \beta \end{aligned}\right\} \tag{6.1.17}$$

There are many ways to find a solution of (6.1.17). (Remember that the question of uniqueness has already been settled - see (6.1.3).) You may use the *method of undetermined coefficients* (see [10] page 117 or [62] page 221 or [58] page 160 or [50] page 135 or [56] page 59), or the *method of variation of parameters* (see [10] page 124 or [62] page 233 or [58] page 168 or [50] page 141 or [57] page 318 or [56] page 72), or *differential operators* (see [62] pages 262 and 288 or [58] page 171), or Laplace transforms. Actually, in this case it is simpler to guess the solution!

Note first of all that if $u(t)$ and $v(t)$ are any two solutions of the inhomogeneous differential equation (6.1.2), then their difference $z(t) = u(t) - v(t)$ is a solution of the homogeneous equation (6.1.4). This is, in fact, true of any linear equation (1.2.1), but we shall formulate the following theorem only for our special case:

Theorem 6.1.1

If the function $u(t)$ is any particular solution of the inhomogeneous differential equation in (6.1.2), then every solution of this equation is the sum of $u(t)$ and a solution of the corresponding homogeneous differential equation in (6.1.4).

If all the solutions of the homogeneous equation are known, we thus need only find one solution of the inhomogeneous equation to obtain all the possible solutions of the inhomogeneous equation! This one solution of the inhomogeneous equation is called *a particular integral.*

Since we have already discussed the solution of the homogeneous case (free motion), and we need only find a particular integral of the differential equation in (6.1.17), we can just as well quickly try a solution of the form

$$A\cos(at+b).$$

Substitute this function in (6.1.17):

$$-ma^2A\cos(at+b)+kA\cos(at+b)=E\cos(at+b)$$

which will hold for all t if

$$(k-ma^2)A=E$$

or equivalently, using (6.1.14),

$$m(w_0^2-a^2)A=E$$

Since $w_0 \neq a$, we can solve for the constant A to obtain the particular integral. Hence, by Theorem 6.1.1 every solution of the differential equation in (6.1.17) is of the form

$$x(t)=d_1\cos w_0t+d_2\sin w_0t+\frac{E}{m(w_0^2-a^2)}\cos(at+b) \tag{6.1.18}$$

An expression which describes all the possible solutions of a differential equation is called *the general solution.* As in the case of the first order linear equation (see §2.1) we shall call the general solution of the homogeneous differential equation *the complementary function.* Thus, the general solution of the differential equation in (6.1.17) is the sum of the complementary function and any particular integral.

The initial values in (6.1.17) must still be satisfied:

$$\left.\begin{aligned}\alpha &= d_1+\frac{E}{m(w_0^2-a^2)}\cos b\\ \beta &= d_2w_0-\frac{Ea}{m(w_0^2-a^2)}\sin b\end{aligned}\right\} \tag{6.1.19}$$

Since $k>0$, and thus $w_0\neq 0$, the constants d_1 and d_2 can now be found. If these values of d_1 and d_2 are substituted in (6.1.18), we finally have the unique solution of (6.1.17).

Note that the general solution can also be expressed as

$$x(t) = d_3 \cos(w_0 t - \delta) + \frac{E}{m(w_0^2 - a^2)} \cos(at + b)$$

with

$$\tan \delta = \frac{d_2}{d_1} \text{ and } d_3 = \sqrt{d_1^2 + d_2^2}$$

This shows that $x(t)$ is the sum of two simple harmonic motions in the same straight line, but with different periods (since $w_0 \neq a$). Note that $x(t)$ is not a simple harmonic motion and is, in fact, not even periodic in general; to be periodic we must have

$$\frac{a}{w_0} = \frac{p}{q} \qquad (p, q \text{ positive integers })$$

in which case the period will be $\leq \frac{2p\pi}{a}$, or equivalently $\leq \frac{2q\pi}{w_0}$. Moreover, the motion is bounded, since the displacement from the position of static equilibrium cannot exceed

$$d_3 + \frac{E}{m \mid w_0^2 - a^2 \mid}.$$

An interesting special case occurs when the bob starts from rest at the position of static equilibrium, and the phase angle of the external force is zero; in symbols, $\alpha = \beta = b = 0$. By (6.1.18) and (6.1.19) we have

$$x(t) = \frac{E}{m(w_0^2 - a^2)}(\cos at - \cos w_0 t)$$

which can be rewritten as

$$x(t) = \frac{2E}{m(w_0^2 - a^2)} \sin(\frac{w_0 - a}{2} t) \sin(\frac{w_0 + a}{2} t) \qquad (6.1.20)$$

by using the trigonometric identity

$$\cos(A - B) - \cos(A + B) = 2 \sin A \sin B.$$

Suppose that $w_0 + a$ is much bigger than $\mid w_0 - a \mid$. Then (6.1.20) is the product of a fast oscillating motion and a slow oscillating motion, as shown in Figure 6.1.3.

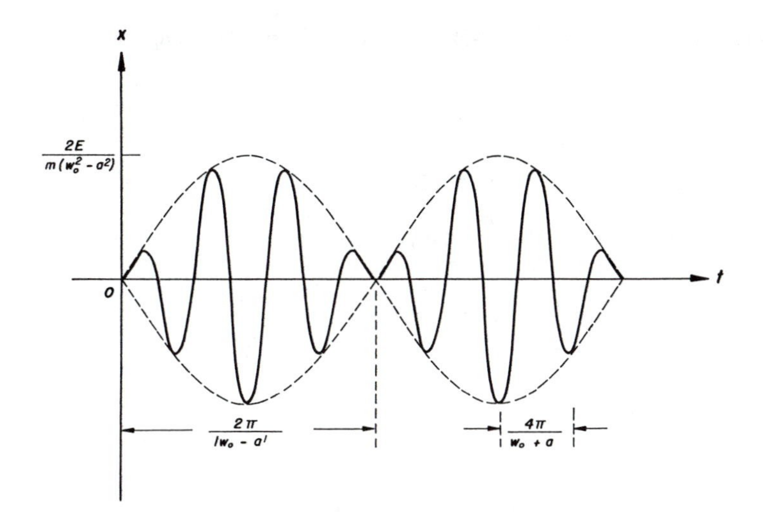

Figure 6.1.3: **Undamped motion with** $w_0 \approx a$

We could also see the graph as a motion with a high frequency

$$\frac{w_0 + a}{4\pi}$$

and an amplitude which changes slowly with the frequency

$$\frac{\mid w_0 - a \mid}{4\pi}.$$

In fact, this behaviour can by demonstrated when two tuning forks with nearly equal frequencies are struck simutaneously - the periodic variation in the loudness (or the amplitude) can clearly be heard.

(b) Resonance: $c = 0, w_0 = a$

The equation (6.1.17) now reads

$$\left.\begin{aligned} x'' + w_0^2 x &= \frac{E}{m}\cos(w_0 t + b) \qquad (0 < t < \infty) \\ x(0) &= \alpha, \\ x'(0) &= \beta \end{aligned}\right\} \qquad (6.1.21)$$

Use Laplace transforms to find the solution, assuming that $x(t)$ is of class E with the transform $X(s)$:

$$s^2 X(s) - s\alpha - \beta + w_0^2 X(s) = \frac{E\cos b}{m}\frac{s}{s^2 + w_0^2} - \frac{E\sin b}{m}\frac{w_0}{s^2 + w_0^2}$$

where we have used Table 4.4.1, numbers 7 and 8. Solve for $X(s)$:

$$X(s) = \frac{s\alpha + \beta}{s^2 + w_0^2} + \frac{E\cos b}{m}\frac{s}{(s^2 + w_0^2)^2} - \frac{E\sin b}{m}\frac{w_0}{(s^2 + w_0^2)^2}$$

Use Table 4.4.1, numbers 7, 8, 13, and 14, to find the inverse transform:

$$\left.\begin{aligned} x(t) &= \alpha\cos w_0 t + \frac{\beta}{w_0}\sin w_0 t + \frac{E\cos b}{2mw_0} t\sin w_0 t \\ &\quad - \frac{E\sin b}{2mw_0^2}(\sin w_0 t - w_0 t\cos w_0 t) \\ &= \alpha\cos w_0 t + (\frac{\beta}{w_0} - \frac{E\sin b}{2mw_0^2})\sin w_0 t + \frac{Et}{2mw_0}\sin(w_0 t + b) \end{aligned}\right\} \tag{6.1.22}$$

This is the unique solution of (6.1.21). The amplitude of the last term grows as the time t increases, with the result that the oscillation is unbounded, as shown in Figure 6.1.4.

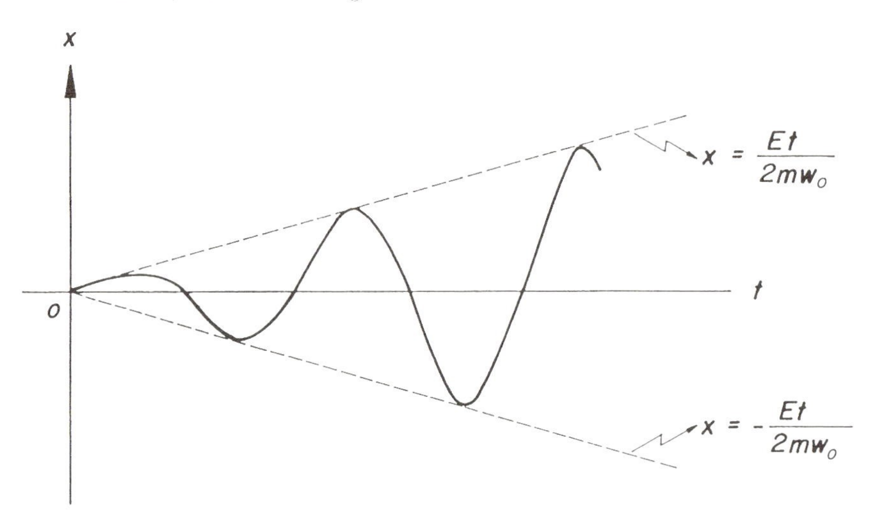

Figure 6.1.4: **Resonance in undamped motion**

This phenomenon of unbounded oscillation when the natural frequency $\frac{w_0}{2\pi}$ coincides with the frequency $\frac{a}{2\pi}$ of the external force is called *resonance*. This could happen when men are marching over a bridge, when waves hit a concrete damwall, or even when wind impacts on a on structure like a bridge or a tower. In actual fact the structure would break up when the oscillations get too large, so that the oscillation never really becomes unbounded.

(c) Damped Motion: $c \neq 0$

We now look at motions where damping cannot be ignored. The corresponding initial value problem is

$$\left.\begin{array}{rcll} mx'' + cx' + kx & = & E\cos(at+b) & (0 < t < \infty) \\ x(0) & = & \alpha, & \\ x'(0) & = & \beta & \end{array}\right\} \quad (6.1.23)$$

The case of the homogeneous equation was already studied in (i)(a) -(c) above. To find a particular integral, let us consider a function of the type

$$x(t) = A\cos(at+b) + B\sin(at+b)$$

where A and B are constants yet to be determined. Substitute in (6.1.23) and collect sine and cosine terms:

$$\begin{array}{rcl} E\cos(at+b) & = & (-ma^2A + caB + kA)\cos(at+b) \\ & & +(-ma^2B - caA + kB)\sin(at+b) \end{array}$$

To ensure that this equation holds for all positive values of t, we must require that

$$\begin{array}{rcl} (k - ma^2)A + caB & = & E \\ -caA + (k - ma^2)B & = & 0 \end{array}$$

For convenience let us define

$$\Delta = (k - ma^2)^2 + c^2a^2 \quad (6.1.24)$$

which is positive, since both c and k are nonzero. Then we obtain

$$A = \frac{E(k - ma^2)}{\Delta} \text{ and } B = \frac{acE}{\Delta}$$

We have then the particular integral

$$x(t) = \frac{E(k - ma^2)}{\Delta}\cos(at+b) + \frac{acE}{\Delta}\sin(at+b) \quad (6.1.25)$$

which can be written in a more compact form by introducing an auxiliary angle θ

$$x(t) = \frac{E}{\sqrt{\Delta}}\cos(at + b - \theta) \quad (6.1.26)$$

where θ is defined by

$$\tan\theta = \frac{ac}{k - ma^2}.$$

Note that θ is an angle between zero and π radians because ac is positive. The general solution of the differential equation in (6.1.23) can now be given as

$$x(t) = z(t) + \frac{E}{\sqrt{\Delta}} \cos(at + b - \theta) \tag{6.1.27}$$

where $z(t)$ is the complementary function which was already found in the cases (i)(a) - (c) for the different types of damping. The two arbitrary constants in $z(t)$ are now determined by the two initial conditions in (6.1.23), which then produces the unique solution of (6.1.23).

Note in all three cases (i)(a) -(c) that for any positive value of c it is always true that $z(t)$ tends exponentially to zero as $t \to \infty$. Hence, after a short while $z(t)$ will be small and the solution $x(t)$ will be close to the expression in (6.1.26). In the literature $z(t)$ is referred to as the *transient vibrations* and (6.1.26) as the *steady-state vibrations*. Note also that θ is the phase difference between the forcing term and the steady-state vibrations.

Broadly speaking, the complementary function, or the transient vibrations, allows that a given initial state can be satisfied; and with increasing time the initial energy due to the displacement and speed is dissipated by the damping so that the motion is, in the end, only the reaction of the system to the external force.

It is interesting to investigate which value of a will maximize the amplitude $E\Delta^{-\frac{1}{2}}$ in (6.1.26) for fixed values of m, c, k, and E. Since E is fixed, we must determine the value of a which will minimize Δ. Differentiate (6.1.24):

$$\frac{d\Delta}{da} = 2(k - ma^2)(-2ma) + 2ac^2$$

and equate to zero to obtain

$$a^2 = \frac{2km - c^2}{2m^2} \tag{6.1.28}$$

If $c^2 < 2km$, then the square root provides a real value of a in (6.1.28) - this value of a will indeed minimize Δ, since the second derivative is positive there. (Note that in this case the motion is underdamped, as could be expected.) The maximal amplitude is then

$$\text{Amplitude} = \frac{2mE}{c\sqrt{4mk - c^2}} = \frac{E}{cw} \tag{6.1.29}$$

with w as defined in (6.1.9).

Perhaps you are wondering of what practical use this maximal amplitude could be! Suppose you are part of a team to design a seismograph which must detect vibrations in the rock layers of a mine. These vibrations play the role of the external force, and since their approximate frequencies are known, a is known. (Note incidentally that $\alpha = \beta = 0$, since the system is at rest when it picks up the vibration.) We shall design the spring so that c is small, and then choose k and m so that (6.1.28) is satisfied. Now we know that the amplitude of the steady-state vibrations will be maximal, and the low value for c will ensure in (6.1.29) that small vibrations (E small) can be detected.

- Read: Any one of [10] page 129, [62] page 347, [58] page 191, [50] page 159, [57] page 299, [56] page 80, [59] Chapter 1. See also [31] pages 17, 40, and 64 for interesting models of price speculation. In [8] a model of traffic flow is discussed, consisting of a system of second order differential equations and solved by means of Laplace transforms.

- Do: Exercises 1, 2, 3 in §6.5.

- Do: Project A^* in §6.6.

6.2 Electrical Networks

Special networks were discussed in §5.5, where each loop in the network was modelled as a first order differential equation. Many networks, however, include loops which will produce second order differential equations. Consider, for example, the loop in Figure 6.2.1.

As in §5.5 we find that

$$L\frac{di}{dt} + Ri - E(t) + \frac{q}{C} = 0 \tag{6.2.1}$$

where i and q denote the current in the loop and the charge on the capacitor, respectively, at any time t. Since by definition (see (5.5.1) in §5.5)

$$i = \frac{dq}{dt} \tag{6.2.2}$$

the equation (6.2.1) can be rewritten as

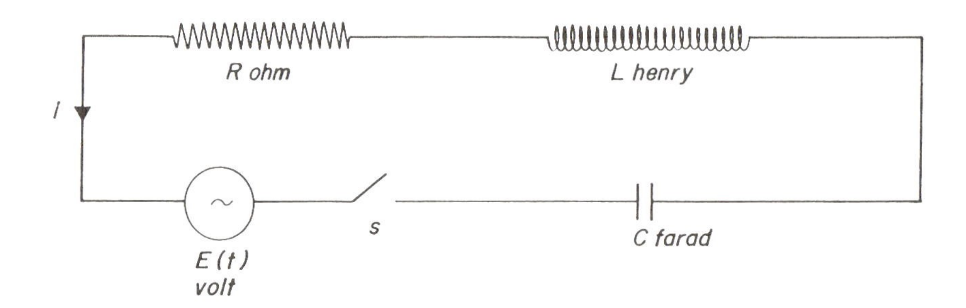

Figure 6.2.1: **Network of (6.2.3)**

$$\left.\begin{aligned} Lq'' + Rq' + \frac{1}{C}q &= E(t) \qquad 0 < t < \infty \\ q(0) &= \alpha, \\ q'(0) &= \beta \end{aligned}\right\} \qquad (6.2.3)$$

Compare now the two initial value problems (6.2.3) and (6.1.2). It is clear that from a mathematical viewpoint the two problems are the same. In fact, the solutions will be identical if we interchange

the displacement $x(t)$	with the charge $q(t)$
the velocity $x'(t)$	with the current $q'(t) = i(t)$
the mass m	with the inductance L
the damping factor c	with the resistance R
the spring constant k	with the reciprocal capacitance $\frac{1}{C}$
the external force $f(t)$	with the voltage source $E(t)$.

As in §6.1, we also use here the terminology

overdamping	when $R^2C > 4L$
critical damping	when $R^2C = 4L$
underdamping	when $R^2C < 4L$

When $E(t) = E\cos(at+b)$, we shall also speak of *resonance* when $R = 0$ and $a = w_0 = (LC)^{-\frac{1}{2}}$, and we shall use the terminology *transient* and *steady-state* solution in the same sense as in §6.1.

The correspondence between the network in Figure 6.2.1 and the initial value problem (6.1.2) is, in fact, the basis on which an analogue computer works. When a system like (6.1.2) must be analysed, the corresponding network is set up, and the charge q and the current i are measured or displayed on an oscillograph.

- Read: Any of [10] page 142, [62] page 369, [58] page 205, [21] page 15, or [50] page 172.

- Do: Exercises 4, 5, 6 in §6.5.

6.3 The Ignition of an Automobile

The purpose of this section is to show how a complicated problem can be broken up into several pieces in such a way that the analysis of each piece is fairly simple.

The ignition system of an automobile consists essentially of a voltage source in the form of a battery or a generator and a transformer connected over the spark plug, as shown in Figure 6.3.1. The transformer has a primary network (consisting of two loops) which contains a switch, a capacitor of C farad, a resistor of R ohm, and the inductance L henry of the transformer. The switch in the primary network is opened and closed mechanically by the rotor in the distributor. The transformer also has a secondary loop which consists of a resistor of $\bar{R}$ ohm, the inductance $\bar{L}$ henry of the transformer, and the small opening between the points of the spark plug. When a spark jumps over the points, the fuel mixture in the cylinder is ignited.

As the switch closes in the primary network, the voltage increases in the secondary loop until a spark flashes over the points of the spark plug, which then closes the secondary loop. Now current flows in the secondary loop, the voltage drops, and the spark vanishes.

To simplify matters, let us first split the process up into four stages, and then attempt to model each stage. Each of these stages is shown in Figure 6.3.1.

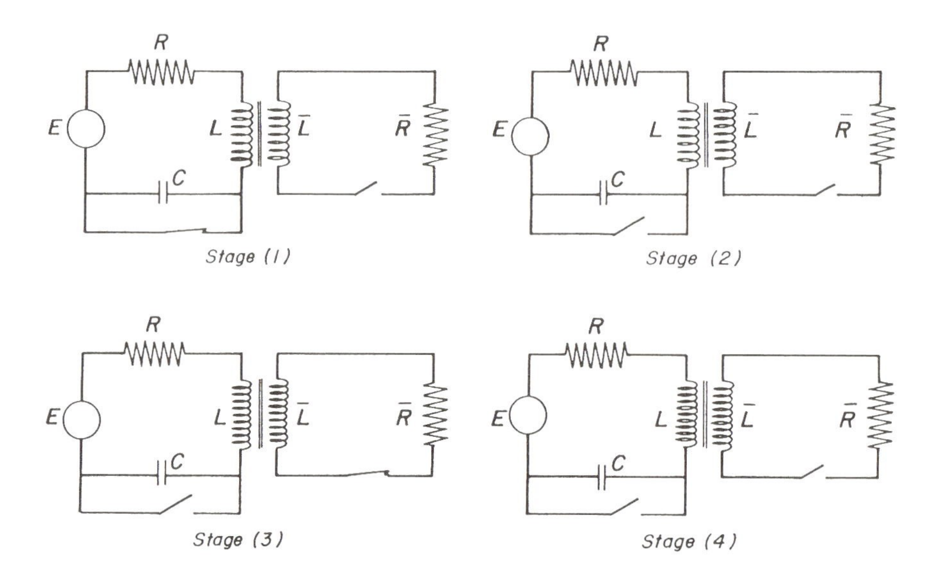

Figure 6.3.1: **The four stages in the ignition cycle**

Stage (1): No current flows in the primary network or the secondary loop when the rotor closes the switch in the primary network at $t = 0$. The current in the primary loop increases with increasing time, but no current flows in the secondary loop.

Stage (2): The rotor opens the switch in the primary loop. The capacitor ensures that the current is cut off cleanly without sparks at the switch and causes an oscillatory current in the rest of the primary network. This fast-changing direction of the current in the primary network affects the magnetic field in the core of the transformer, which in turn induces a voltage over the part of the transformer in the secondary loop.

Stage (3): The voltage in the secondary loop is high enough so that a spark appears in the spark plug, with the result that current flows in the secondary loop. In the primary network the oscillating current is still flowing.

Stage (4): The voltage drops in the secondary loop, the spark disappears, and the current stops. In the primary network the resistor causes the oscillating current to weaken and finally to die off.

We are now ready to model these four stages. Let us denote the currents in the primary network and the secondary loop by $i(t)$ and $\bar{i}(t)$, respectively.

(1) *The current increases in the primary network.*

At $t = 0$ when the switch closes, current starts to flow in the outer loop of the primary network. By Kirchhoff's laws we obtain the initial value problem

$$L\frac{di}{dt} + Ri = E \qquad i(0) = 0 \tag{6.3.1}$$

This is a linear differential equation with the integrating factor $e^{\frac{Rt}{L}}$ (see §2.1). After integration we find that the unique solution is

$$i(t) = \frac{E}{R}(1 - e^{\frac{-Rt}{L}}) \tag{6.3.2}$$

(2) *The voltage increases in the secondary loop.*

As the switch opens the current flows to the capacitor. Let $q(t)$ denote the charge on the plates of the capacitor, and let $t = 0$ be the moment when the switch had opened. Just as in §6.2, we now find that

$$\left.\begin{array}{rcl} Lq'' + Rq' + \frac{1}{C}q & = & E \\ q(0) = 0, \; q'0) & = & \beta \end{array}\right\} \tag{6.3.3}$$

where β represents the current at the moment when the switch opens. We can, in fact, find β because we must have by (6.3.2) that

$$\beta = \frac{E}{R}(1 - e^{\frac{-Rt_1}{L}}) \tag{6.3.4}$$

where t_1 is the time during which the switch was closed. In a typical ignition system we would have C between 10^{-7} and 5×10^{-7} farad, R between 1 and 2 ohm, and L between 5×10^{-3} and 10^{-2} henry. This implies that $R^2C < 4L$, and thus we always have underdamping.

To solve (6.3.3), let us first tackle the complementary function. As in §6.1, we use the auxiliary equation (which results when a function of the form $e^{\lambda t}$ is substituted in the homogeneous equation):

$$L\lambda^2 + R\lambda + \frac{1}{C} = 0.$$

Since $R^2C < 4L$, define the positive real number

$$w = \sqrt{\frac{1}{LC} - \frac{R^2}{4L^2}} \tag{6.3.5}$$

so that the complex conjugate roots of the quadratic equation in λ can be written

$$\lambda = -\frac{R}{2L} \pm iw$$

and hence (see (6.1.9) in §6.1) the complementary function is

$$q(t) = e^{-\frac{Rt}{2L}}(A\cos wt + B\sin wt).$$

To find a particular integral of (6.3.3) let us try a constant, since E is a constant. Let

$$q(t) = \gamma \qquad (\gamma \text{ is a constant})$$

and substitute into the differential equation to obtain $\gamma = CE$.

Hence, the general solution is

$$q(t) = e^{-\frac{Rt}{2L}}(A\cos wt + B\sin wt) + CE.$$

The initial values require that

$$0 = A + CE \text{ and } \beta = -\frac{R}{2L}A + wB,$$

and so the unique solution of (6.3.3) is

$$q(t) = CE(1 - e^{-\frac{Rt}{2L}}\cos wt) + e^{-\frac{Rt}{2L}}(\frac{\beta}{w} - \frac{RCE}{2Lw})\sin wt \quad (6.3.6)$$

with β and w given by (6.3.4) and (6.3.5), respectively.

It is, however, the current $i(t)$ in the primary network that we need to know. Differentiate (6.3.6) and use (6.3.5) to simplify:

$$i(t) = e^{-\frac{Rt}{2L}}[\cos wt + \frac{1}{Lw}(\frac{E}{\beta} - \frac{R}{2})\sin wt] \quad (6.3.7)$$

Note in (6.3.7) firstly that w is a large number (in the order of 10^5 according to the values given at (6.3.4) for R, C, and L), with the result that the oscillations of the current in the primary network are very fast; and secondly that the factor $\frac{R}{2L}$ in the exponent is in the order of 10^2, which means that the current tends very quickly to zero.

The oscillatory current in the primary network creates increasing voltage in the secondary loop. In a transformer the voltage in the secondary coil is N times the voltage in the primary coil, where N is the ratio between the turns in the secondary coil to the turns in the primary coil. In our situation N would be between 50 and 100. By (6.3.7) we have

Voltage in the secondary coil

$$\begin{aligned} &= NLi' \\ &= Ne^{-\frac{Rt}{2L}}[(E - R\beta)\cos wt + \frac{R^2\beta - 2RE - 4L^2w^2}{4Lw}\sin wt] \\ &= Ne^{-\frac{Rt}{2L}}G\cos(wt + \phi) \end{aligned}$$

where the phase angle ϕ is introduced in the same way as in (6.1.12) in §6.1. The coefficient of $\sin wt$ is certainly negative (being dominated by Lw), and the coefficient of $\cos wt$ is positive by (6.3.4); hence, ϕ must be a positive acute angle. A rough sketch of the graph of $\cos(wt+\phi)$ will show that the voltage in the secondary coil reaches its peak level in less than $\frac{\pi}{w}$ seconds! This means also that the spark must flash in the secondary loop in less than $\frac{\pi}{w}$ seconds. Hence, the time t_2 for stage (2) is less than $\frac{\pi}{w}$ seconds, which is in the order of 10^{-5} seconds.

(3) *The current flows in the secondary loop.*

While the spark flashes, a current $\bar{i}(t)$ flows in the secondary loop for this small time interval. Depending on the gap in the spark plug, the time interval is about 10^{-5} seconds.

The currents in the primary network and secondary loop are described by the simultaneous differential equations

$$\left.\begin{aligned} Li' + M\bar{i}' + Ri + \tfrac{1}{C}q &= E \\ \bar{L}\bar{i}' + \bar{R}\bar{i} &= Mi' \end{aligned}\right\} \qquad (6.3.8)$$

where M denotes the mutual inductance between the primary and the secondary coils. We shall not solve these equations here because the time interval concerned is so short. (See, however, Exercise 7!)

(4) *The current dies off in the primary network.*

When the spark stops, the current in the secondary loop stops and the primary network reverts to the state just before the spark. So (6.3.3) models the situation, except that the initial values will be different. Hence, the current in the primary network oscillates fast with a sharply decaying amplitude due to the exponential factor in the solution.

We have analysed the four stages in the ignition cycle. Next we would like to draw a graph of this cycle. In the engine of an automobile the rotor which opens and closes the primary network is driven at half the revolutions of the crankshaft. For example, if the crankshaft is turning at 3600 revolutions per minute, then the rotor completes one revolution in $\frac{1}{30}$ second. In a four-cylinder engine where four spark plugs must fire in one revolution of the rotor, only about 20° of the revolution is available to close the primary network of one spark plug. Thus the time interval when the primary network is closed is about

$$\frac{20}{360} \times \frac{1}{30} = \frac{1}{540} = 0.00185 \text{ second}$$

For the rest of the revolution (0.03333 - 0.00185 = 0.03148 second) the primary network is open. Since each of the stages (2) and (3) took only about 10^{-5} seconds their time interval is far smaller than the other two stages. In the rough graph of the current $i(t)$ shown in Figure 6.3.2, the stages (2) and (3) appear to be at the same time instant.

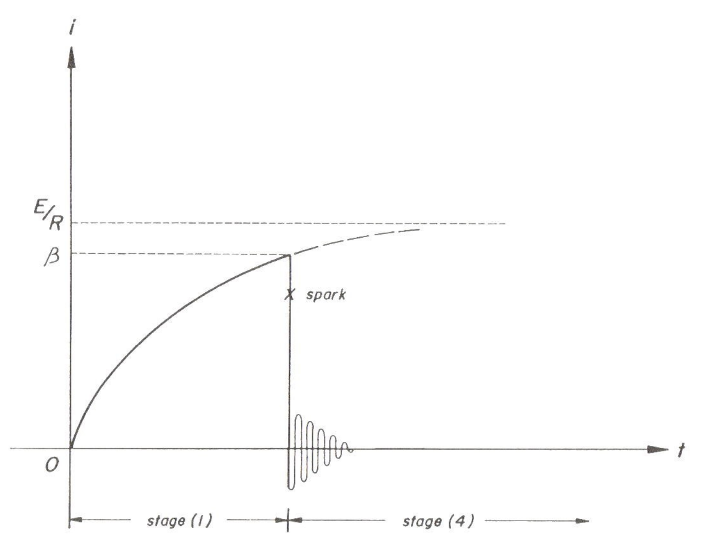

Figure 6.3.2: **Current in the primary network**

Note that the time interval t_1, when the primary network is closed, decreases if the revolutions of the engine increase. When t_1 is smaller, the peak current in the primary network is less, with the result that the voltage in the secondary loop might not be high enough to cause a spark. In modern automobiles whose engines run at high revolutions and at high compression, this problem is acute. It was solved by introducing a 12-volt battery instead of the old 6-volt battery, and by doubling the resistance R of the resistor. Thus the current $i(t)$ still tends to the same limiting value $\frac{E}{R}$, but the same β of (6.3.4) is reached in a shorter time.

- Read: [46] page 93 where an ignition system is also discussed.
- Do: Exercises 7, 8 in §6.5.

6.4 Simultaneous Equations

In Chapter 5 we have solved systems of two first order linear differential equations. We could also eliminate one of the dependent variables to obtain a second order differential equation. Consider, for example, the system (5.9.1) of Project C:

$$\begin{aligned} x' &= -bx + ay, \qquad & x(0) &= \alpha \\ y' &= bcx, & y(0) &= \beta \end{aligned}$$

By the first equation we have

$$y = \tfrac{1}{a}(x' + bx) \tag{6.4.1}$$

which we can differentiate

$$y' = \tfrac{1}{a}(x'' + bx')$$

and substitute in the second equation to obtain

$$x'' + bx' - abcx = 0. \tag{6.4.2}$$

The initial values are of course $x(0) = \alpha$ and by (6.4.1)

$$\frac{1}{a}(x'(0) + bx(0)) = \beta$$

which can be rewritten as

$$x'(0) = a\beta - b\alpha \tag{6.4.3}$$

We must now solve the second order differential equation (6.4.2) with the initial values $x(0) = \alpha$ and (6.4.3). Try a solution of the type $e^{\lambda t}$ to obtain the auxiliary equation

$$\lambda^2 + b\lambda - abc = 0$$

Let p and q be the roots of this equation:

$$\begin{aligned} p &= \frac{-b + \sqrt{b^2 + 4abc}}{2} \\ q &= \frac{-b - \sqrt{b^2 + 4abc}}{2} \end{aligned}$$

then the general solution is given by

$$x(t) = Ae^{pt} + Ce^{qt} \tag{6.4.4}$$

The constants A and C are determined by the initial conditions:

$$\begin{aligned} \alpha &= A + C \\ a\beta - b\alpha &= pA + qC \end{aligned}$$

When these values of A and C are substituted in (6.4.4) we obtain the same function for $x(t)$ as in (5.9.2). To find $y(t)$ we need only substitute $x(t)$ in (6.4.1).

We can use the same technique for more complicated situations. Consider, for example, the system (6.3.8):

$$\begin{aligned} Li' + M\bar{i}' + Ri + \tfrac{1}{C}q &= E \\ \bar{L}\bar{i}' + \bar{R}\bar{i} &= Mi' \end{aligned}$$

Use the operator notation $D = \frac{d}{dt}$ to rewrite these equations as

$$(LD + R)i + MD\bar{i} + \tfrac{1}{C}q \quad = \quad E \tag{6.4.5}$$

$$(\bar{L}D + \bar{R})\bar{i} \quad = \quad MDi. \tag{6.4.6}$$

We first eliminate $\bar{i}$. Differentiate (6.4.5) and multiply by $\bar{L}$ (or equivalently, multiply by $\bar{L}D$), and add the result to (6.4.5) multiplied by $\bar{R}$:

$$(\bar{L}D + \bar{R})(LD + R)i + MD(\bar{L}D + \bar{R})\bar{i} + \tfrac{1}{C}(\bar{L}D + \bar{R})q = \bar{R}E$$

Now use (6.4.6) in the second term:

$$(\bar{L}D + \bar{R})(LD + R)i + MDMDi + \tfrac{1}{C}(\bar{L}D + \bar{R})q = \bar{R}E$$

and using $Dq = i$ to simplify

$$(\bar{L}LD^2 + \bar{R}LD + \bar{L}RD + \bar{R}R)i + M^2D^2i + \tfrac{1}{C}\bar{L}i + \tfrac{1}{C}\bar{R}q = \bar{R}E$$

we have succeeded in eliminating $\bar{i}$.

Next we eliminate q by differentiating once more:

$$(\bar{L}LD^3 + \bar{R}LD^2 + \bar{L}RD^2 + \bar{R}RD)i + M^2D^3i + \tfrac{1}{C}\bar{L}Di + \tfrac{1}{C}\bar{R}i = 0$$

and thus obtain the third order differential equation

$$(\bar{L}L + M^2)i''' + (\bar{R}L + \bar{L}R)i'' + (\bar{R}R + \tfrac{1}{C}\bar{L})i' + \tfrac{1}{C}\bar{R}i = 0 \tag{6.4.7}$$

which can be solved in the same way as the linear second order differential equation with constant coefficients. For example, consider the auxiliary equation (see after (6.1.4) in §6.1):

$$(\bar{L}L + M^2)\lambda^3 + (\bar{R}L + \bar{L}R)\lambda^2 + (\bar{R}R + \tfrac{1}{C}\bar{L})\lambda + \tfrac{1}{C}\bar{R} = 0 \tag{6.4.8}$$

A cubic polynomial has at least one real root, and since all the coefficients of the polynomial are positive, any real root must be negative to obtain zero on the right hand side of (6.4.8). Moreover, when this characteristic polynomial is differentiated a quadratic polynomial is obtained. The points on the graph of the cubic polynomial in (6.4.8), where the slope is zero, can be found - note that these values of λ must also be negative when they are real numbers. Since λ appear in the exponent of the solution, it follows that if all three roots are negative or have negative real parts, then the solution will tend to zero as $t \to \infty$. This aspect is investigated in §6.5, Exercise 10.

Given numerical values for the parameters, a suitable numerical algorithm can be utilized to determine the three roots of (6.4.8), as was discussed in §1.4 in the first chapter. When these values are substituted in the complementary function and appropriate boundary values are used to determine the arbitrary constants, the solution $i(t)$ of (6.4.7) is obtained. It is now a simple matter to substitute $i(t)$ in (6.4.6) and to solve this linear first order equation (see §2.1) for $\bar{i}(t)$. Finally, we can use (6.4.5) to find $q(t)$.

We have shown in this section that a system of linear differential equations can be written as a single differential equation of higher order. There are no hard and fast rules whether to solve a problem as a system of first order differential equations or as a single higher order equation. It would depend on the specific problem and the method utilized for the solution.

- Read: [10] page 266 or [36] page 91 or [50] page 226 or [58] page 215 or [62] page 400.
- Do: Exercises 9, 10, 11 in §6.5.
- Do: Project B in §6.6.

6.5 Exercises

(1) Use Laplace transforms to solve (6.1.4). Compare your answer with (6.1.6), (6.1.8), and (6.1.10).

(2) In (6.1.2) let the external force be

$$f(t) = \begin{cases} a & (0 \le t \le b) \\ 0 & (t > b) \end{cases}$$

with a and b constants. This models the situation where an impulsive force acts on the bob, for example, when the bob receives a hard knock downwards. Suppose that $\alpha = \beta = 0$ and that

$$k = \frac{m^2 + c^2}{4m}.$$

Solve (6.1.2) for this special case.

(3) Solve (6.1.2) for the special case when $c = 0, m = 4k$ and $f(t)$, as in Exercise 2. If $\alpha = \beta = 0$, show that for $t \geq b$

$$x(t) = \frac{2a}{k} \sin\frac{b}{4} \sin\frac{2t - b}{4}.$$

Use this expression to find the amplitude and the period of the motion $x(t)$. Draw a rough graph of $x(t)$ for $t \geq 0$.

(4) A loop consists of an inductor L, a resistor R, a capacitor C, and a switch. When the switch is closed, the capacitor has a charge of 10^{-5} coulomb. Find the charge at any time t in each of the following cases:

(a) $L = 0.2$ henry, $C = 10^{-5}$ farad, $R = 300$ ohm;

(b) $L = 1.0$ henry, $C = 4 \times 10^{-6}$ farad, $R = 1000$ ohm;

(c) $L = 2.0$ henry, $C = 10^{-5}$ farad, $R = 400$ ohm.

(5) A loop consists of an inductor of 1 henry, a resistor of 5000 ohm, a capacitor of 0.25×10^{-6} farad, a battery of 12 volt, and a switch. When the switch is closed, the capacitor has a zero charge. Find the charge $q(t)$ coulomb when (i) $t = 0.001$ seconds and (ii) $t = 0.01$ seconds. What is the limit value of the charge when $t \to \infty$?

(6) Find the steady-state current in a loop consisting of an inductor of 10 henry, a resistor of 3000 ohm, a capacitor of 2.5×10^{-6} farad, and a voltage source of $110 \cos(120\pi t)$ volt.

(7) Write (6.3.8) as a system of three first order equations of the form (5.1.1). Try to solve the system by Laplace transforms. (You will end up with a cubic in s which cannot be factored readily. We shall discuss this cubic in §6.4 - see (6.4.8).)

(8) Use the program OURGRAPH or any other suitable program to draw the graphs of Stage (1) in the model of §6.3 for the two cases

(a) $E = 6$ volt, $R = 1$ ohm, $L = 10^{-2}$ henry;

(b) $E = 12$ volt, $R = 2$ ohm, $L = 10^{-2}$ henry;

on the same axes. Compare the values of β at $t = t_1 = 0.00185$.

(9) Investigate whether a solution exists for the system

$$\begin{aligned} 2x' - 2y' &= y \\ -x' + y' &= -2x \\ x(0) &= 1 \\ y(0) &= 2. \end{aligned}$$

Can the situation be changed if some other value is assigned to $y(0)$?

(10) Given any cubic polynomial with positive coefficients, deduce a condition on the coefficients to ensure that all the roots of the polynomial have negative real parts. Use your result to investigate if this is indeed the case in (6.4.8).

(11) Find the steady-state current $i(t)$ in the network shown in Figure 6.5.1 below.

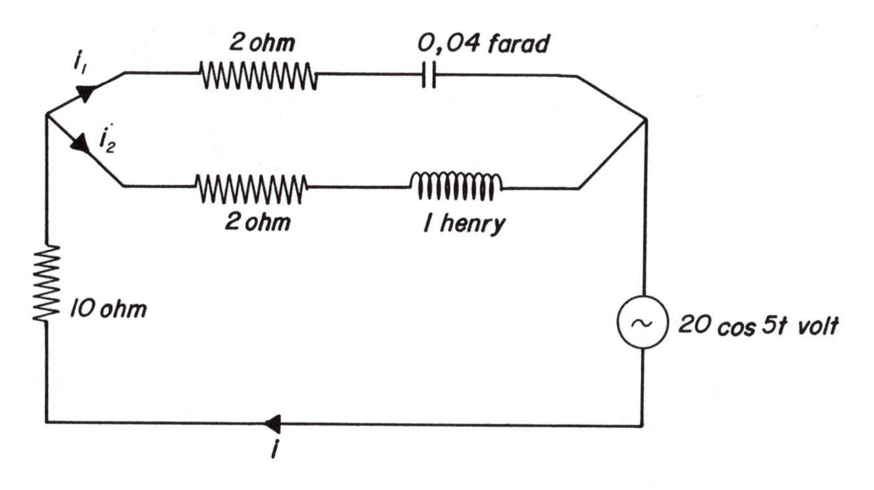

Figure 6.5.1: **Network of Exercise 11.**

6.6 Projects

Project A^*: Motion of a Spring

When bob of mass 0.125 kilograms is attached to a long spring hanging from a hook, the spring extends another 4.9 centimeter to the position of static equilibrium. From this position the spring is pulled down 3.5

centimeters and then released. Suppose that underdamping with damping factor c kg/s occurs and that Assumption (U) of §6.1 holds. Choose an appropriate x-axis for the displacement of the bob.

(1) Construct a model to determine the motion $x(t)$ of the bob where t denotes the time in seconds. (Do not use the formulas in §6.1 - start from scratch!)

(2) Suppose that the bob passes through the origin every $\frac{\pi}{14}$ seconds. Use this information to calculate c.

(3) Use the program OURGRAPH in §1.3 (or any other suitable program of your choice) to draw the graph of $x(t)$ for $0 \leq t \leq 2$.

(4) Suppose an external force $f(t) = 0.1 \cos 10t$ acts on the bob from the beginning of the motion. Find the solution $x(t)$ in this case. (Do not use the formulas in §6.1!)

(5) Use the computer program in (3) above to draw the graphs of $x(t)$ and $f(t)$ of (4) above on the same axes for $0 \leq t \leq 3$.

Project B^*: Motion of Double Springs

Suppose two bobs with mass m_1 and m_2, respectively, are attached to two springs with spring constants k_1 and k_2, respectively, as shown in Figure 6.6.1. Let the displacement of the bob with mass m_1 be denoted by $x(t)$ and the bob with mass m_2 be denoted by $y(t)$ at any time t where x and y are measured downward from the position of static equilibrium at time $t = 0$.

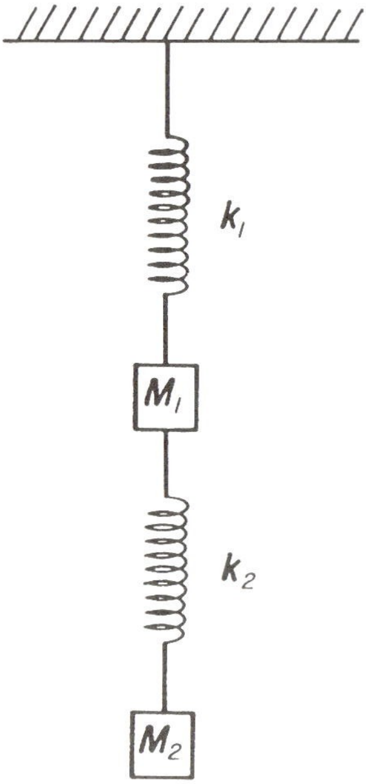

Figure 6.6.1:
System of two springs

(a) Ignore air resistance to construct a model for the motion of the two bobs. Show that the motion is determined by the system of differential equations

$$\begin{aligned} m_2 y'' &= -k_2(y-x) \\ m_1 x'' &= -k_2(y-x) - k_1 x \end{aligned}$$

(b) Suppose the bob with mass m_1 is pulled downward over a distance α and, in addition, the bob with mass m_2 is pulled downward over a distance β. From this position both bobs are released simultaneously at time $t = 0$. Show how you will solve this initial value problem. (Do not write the solution in detail, unless you really like complicated expressions!) Describe qualitatively the motion of the bobs.

(c) Suppose $m_1 = 3$ kilograms, $m_2 = 2$ kilograms, $k_1 = k_2 = 60$ newton/meter, $\alpha = 30$ centimeters and $\beta = 0$. Determine the motion of the two bobs in this case.

6.7 Mathematical Background

Consider a function of the type

$$H(t) = \frac{f(t)}{g(t)}$$

where f and g are both either zero or unbounded as $t \to a$. The limit of $H(t)$ is then unknown, and we refer to it as an *indeterminate form*. Several theorems which give information about the limit of H are collectively known as *L'Hôpital's rule*. An excellent discussion of these theorems (and a short note on who L'Hôpital was) can be found in [3] pages 392 - 409.

For our purposes the following theorem will suffice:

Theorem 6.7.1

Assume f and g have derivatives $f'(t)$ and $g'(t)$ at each point of an open interval (a, ∞), and suppose that $\lim_{t\to\infty} f(t) = +\infty$ and $\lim_{t\to\infty} g(t) = +\infty$. Assume also that $g'(t) \neq 0$ for each $t \in (a, \infty)$. If $\frac{f'(t)}{g'(t)} \to L$ as $t \to \infty$ where L is a finite real number, then $H(t)$ also tends to L as $t \to \infty$.

As an example, consider

$$\lim_{t\to\infty} te^{-t}$$

Since $te^{-t} = \frac{t}{e^t}$ and $f(t) = t$ and $g(t) = e^t$ clearly satisfy the conditions of Theorem 6.7.1, we find that

$$\lim_{t\to\infty} te^{-t} = \lim_{t\to\infty} \frac{1}{e^t} = 0$$

7

Second Order Nonlinear Differential Equations

7.1 Introduction

As we had seen in Chapter 6, the general process for obtaining the solution of a linear differential equation can be broken up into smaller, simpler stages: firstly, the complementary function is determined by looking only at the homogeneous equation; secondly, a particular integral of the inhomogeneous equation must be found; thirdly, the boundary values are used to calculate the arbitrary constants in the general solution.

Unfortunately, this very nice scheme does not work when the differential equation is nonlinear. The reason is, of course, that a linear combination of two solutions of a nonlinear differential equation does not necessarily produce another solution. For example, each of the functions

$$x(t) = e^t \text{ and } x(t) = e^{-t}$$

satisfies the nonlinear differential equation

$$(\frac{dx}{dt})^2 - x^2 = 0 \qquad (7.1.1)$$

but their sum

$$x(t) = e^t + e^{-t}$$

does not, since

$$(e^t - e^{-t})^2 - (e^t + e^{-t})^2 = -4 \neq 0.$$

Moreover, the method of Laplace transforms also cannot be used, because we do not have a nice formula for the Laplace transform of a product of two functions. (See the note after Example 6 in §4.3.)

A popular approach to a nonlinear differential equation is to ignore the nonlinear terms and to solve the resulting linear equation. Hopefully, the solution of the linear problem will be close to the solution of the nonlinear problem, if the nonlinear terms are relatively small - but this is not generally true. For example, consider the nonlinear initial value problem of §2.4

$$\frac{dN}{dt} = bN - sN^2, \qquad N(0) = \alpha$$

with the unique solution

$$N(t) = \frac{b\alpha}{s\alpha + (b - s\alpha)e^{-bt}} \tag{7.1.2}$$

Should we decide to ignore the nonlinear term sN^2 on the grounds that the parameter s is small, we obtain the Malthus model as in (2.2.2)

$$\frac{dN}{dt} = bN, \qquad N(0) = \alpha$$

with the unique solution

$$N(t) = \alpha e^{bt} \tag{7.1.3}$$

No matter how small the parameter s is, the functions (7.1.2) and (7.1.3) are very different - in fact, for large t the function in (7.1.2) tends to a limit $\frac{b}{s}$ whilst the function in (7.1.3) becomes unbounded!

For certain types of nonlinear equations mathematicians had devised means of obtaining solutions: for example, the Bernoulli equation (see [50] page 58 or [62] page 95), the Ricatti equation (see [18] page 57 or [36] page 46), or any of the nonlinear equations which we had solved in Chapter 2. In general, however, we must be satisfied with, at most, a qualitative description of the solution and/or numerical results by way of some appropriate algorithm on a computer, as we had seen in Chapter 3. Before we can implement a numerical procedure, we must be certain that a solution does exist and that the solution is unique. (Read again §3.2!) By writing the differential equation as a system of first order differential equations, we can implement a theorem like Theorem 3.2.1 to settle the question of existence and uniqueness. (This was demonstrated in §5.1.) We shall not discuss numerical solutions for systems of differential equations here, since the basic ideas were already studied in Chapter 3, and in practice one would use an appropriate software package.

In this chapter we shall concentrate on the *qualitative behaviour* of the solution of a nonlinear second order differential equation in which the independent variable t does not appear in any of the coefficients of the equation. Let us define this type of differential equation more generally as follows:

Definition 7.1

The system of first order equations

$$\frac{d\boldsymbol{u}}{dt} = \boldsymbol{F}(\boldsymbol{u}) \qquad (7.1.4)$$

in which the functions $\boldsymbol{F}$ *do not depend on the independent variable* t *is called an* autonomous system.

The vector notation of a system of equations in (7.1.4) was discussed in §5.1. Any second order differential equation of the form

$$\frac{d^2x}{dt^2} = f(\frac{dx}{dt}, x) \qquad (7.1.5)$$

can be written in the form (7.1.4) by introducing a new dependent variable v:

$$\left.\begin{aligned} \frac{dx}{dt} &= v \\ \frac{dv}{dt} &= f(v, x) \end{aligned}\right\} \qquad (7.1.6)$$

In this chapter we shall discuss two autonomous systems: a problem in mechanics in the form (7.1.5) and a problem in population dynamics in the form (7.1.4). The tool that we shall use to obtain qualitative information about the solution is known as a *v-x diagram* or the *phase plane*. The idea is to draw a graph on the (v, x)-axes for (7.1.6), or (u_1, u_2)-axes for (7.1.4), in which the independent variable t plays the role of a parameter. These graphs are given various names depending on the type of application: streamlines, paths, orbits, or trajectories. The important thing is that these graphs supply valuable qualitative information about the solution.

- Do: Exercise 1 in §7.5.

7.2 The Pendulum without Damping

Consider a pendulum which is attached to a fixed point O at one end, and is free to rotate about O in a vertical plane. The pendulum consists of a thin, rigid rod with a heavy bob at the lower end. We want to describe the motion of this pendulum.

Having *identified* the problem, let us turn to the next stage of *assumptions*. To simplify the mechanics of the problem, we shall consider the case of a *simple pendulum* by making the following assumption:

> **Assumption (V)**
>
> *The mass of the bob is so much larger than the mass of the thin rod that the pendulum can be modelled as a massless rod with a particle at one end containing all the mass of the pendulum.*

The pendulum and the corresponding simple pendulum are shown in Figure 7.2.1. Let the mass of the bob (or particle) be m kilograms and the length of the simple pendulum be ℓ meters. When the bob is pulled to one side and released, the bob is constrained to move along the circumference of a circle of radius ℓ as it oscillates around the position of equilibrium.

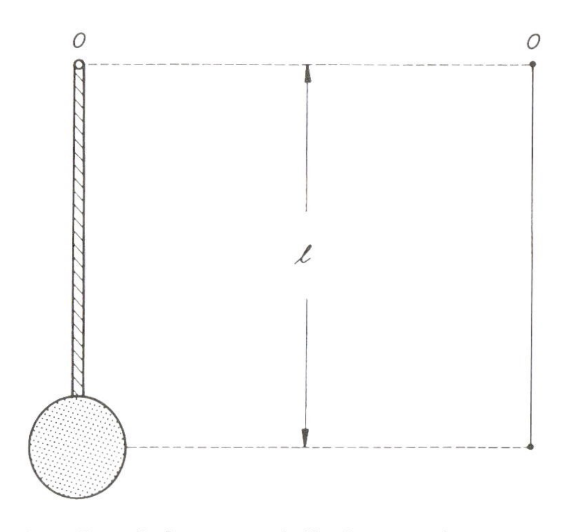

Figure 7.2.1: **Pendulum modelled as a simple pendulum**

Next we look at the *construction stage* of the modelling process. To describe the motion of the pendulum, let $\theta(t)$ be the angular displacement of the rod from the position of equilibrium (shown as OA in Figure 7.2.2) at any time t. We shall measure the angle θ as positive when the bob is to the right of OA and negative when the bob is to the left of OA. Let s be the displacement along the circular arc from A (the position of equilibrium) to P (the position of the bob at time t). Since θ is measured in radians, we know that

$$s = \ell\theta \tag{7.2.1}$$

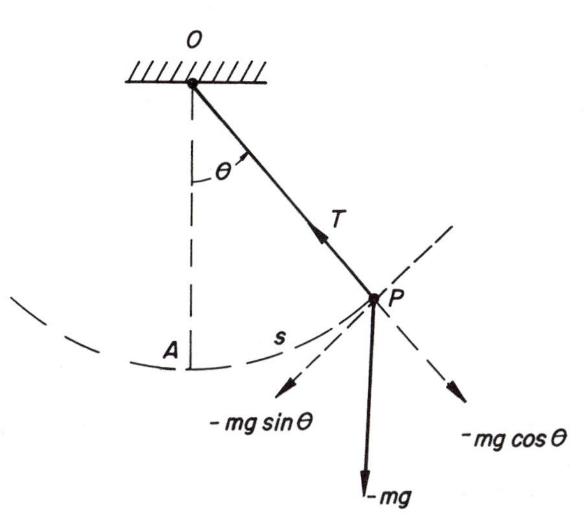

Figure 7.2.2: **Forces acting on the bob**

We shall use Newton's second law (see §2.9) to construct a mathematical model of the motion of the bob. Let g be the acceleration of gravity. Since we ignore air resistance and friction at O (no damping), there are only two forces acting on the bob during the motion. One is the weight of the bob with magnitude mg newtons and direction downwards, and the other is the tension in the rod with magnitude T newtons and direction along the rod, as shown in Figure 7.2.2. To find the resultant force on the bob in the tangential direction, we must resolve the weight into two components, as shown in Figure 7.2.2.

By Newton's second law in the tangential direction we have

$$m\frac{d^2s}{dt^2} = -mg\sin\theta \tag{7.2.2}$$

and differentiating (7.2.1) twice with respect to t (remember that ℓ is constant), we obtain for $0 < t < \infty$ the second order differential equation

$$\theta''(t) = -w^2 \sin\theta(t) \tag{7.2.3}$$

where a prime $'$ denotes the derivative with respect to t, and the parameter w is given by

$$w = \sqrt{\frac{g}{\ell}} \tag{7.2.4}$$

It is interesting to note that had the pendulum been a rigid body of a shape and mass distribution which cannot be modelled as a simple pendulum, we would have used more complicated mechanics of rotating bodies (which falls outside the scope of this book), but in the end would have obtained the same equation (7.2.3), with only a different parameter w. (If you happen to know the mechanical terms, we have in this case that $w^2 = ghk^{-2}$, where h is the distance from O to the center of gravity and k is the radius of gyration of the body.)

Let the initial position and angular velocity of the pendulum be, respectively,

$$\theta(0) = \alpha \quad \text{and} \quad \theta'(0) = \beta \tag{7.2.5}$$

The nonlinear initial value problem (7.2.3), (7.2.5) describes the motion of the simple pendulum, and so the construction stage of the mathematical model is settled. We now turn to the *analysis* of the model.

As was discussed in §7.1, we can introduce a new dependent variable $v(t)$ to rewrite the second order differential equation (7.2.3) as a system of first order differential equations:

$$\left.\begin{aligned} \frac{d\theta}{dt} &= v \\ \frac{dv}{dt} &= -w^2 \sin\theta \end{aligned}\right\} \tag{7.2.6}$$

This system is nonlinear, due to the sin-function, and cannot be solved explicitly to obtain $\theta = f(t)$ and $v = g(t)$, with f and g in terms of elementary functions. But there is an important trick that we can use to obtain qualitative information of these solutions:

If we could eliminate the variable t between the equations $\theta = f(t)$ and $v = g(t)$, we would have

$$F(v, \theta) = 0 \tag{7.2.7}$$

where F is some function of two variables. The graph of F on (v, θ)-axes would then supply some information about the relationship between v and θ. On this graph the time t will play the role of a parameter in the sense that a given value of t will define a point $(\theta(t), v(t))$ on the graph. But how can we do this if we do not know the functions f and g?

By the chain rule of differentiation we have

$$\frac{dv}{d\theta} = \frac{v'(t)}{\theta'(t)} = \frac{-w^2 \sin\theta}{v} \tag{7.2.8}$$

This first order differential equation is separable (see §2.1) and can thus be solved:

$$\int_\beta^v \bar{v}d\bar{v} = -w^2 \int_\alpha^\theta \sin\bar{\theta}d\bar{\theta}$$

$$\frac{1}{2}v^2 - \frac{1}{2}\beta^2 = w^2(\cos\theta - \cos\alpha)$$

This equation can be simplified to

$$v^2 - 2w^2\cos\theta = a \tag{7.2.9}$$

with the constant a representing the value of $v^2 - 2w^2\cos\theta$ at $t = 0$. Thus we found in (7.2.9) the function F of (7.2.7) without knowing what f and g are!

To draw a rough graph of (7.2.9) we note immediately that the graph must be symmetric with respect to the θ-axis because of the fact that v appears only as a square in (7.2.9). Likewise, since the cos-function is an even function, the graph will also be symmetric with respect to the v-axis. The shape of the graph depends on the constant a, or equivalently, on the initial conditions (7.2.5). We must consider five cases separately, namely (i) $a < -2w^2$; (ii) $a = -2w^2$; (iii) $-2w^2 < a < 2w^2$; (iv) $a = 2w^2$; and (v) $a > 2w^2$.

Case (i): $a < -2w^2$

By (7.2.9) we have

$$v^2 < 2w^2\cos\theta - 2w^2 \leq 0$$

since $\cos\theta$ is, at most, 1. Hence, there are no real values of v for any real value of θ; in other words, there is no graph in this case!

Case (ii): $a = -2w^2$

We now have

$$v^2 = 2w^2\cos\theta - 2w^2$$

The only points in the phase plane which can satisfy this equation are $(2n\pi, 0)$ where n is any integer. Hence, the graph consists of isolated points on the θ-axis.

Case (iii): $-2w^2 < a < 2w^2$

In the first quadrant the equation (7.2.9) can now be written

$$v = \sqrt{2w^2\cos\theta + a}. \tag{7.2.10}$$

Since the function $\cos\theta$ is a maximum at $\theta = 0$, this value of $\sqrt{2w^2 + a}$ on the v-axis is also a maximum (see point A in Figure 7.2.3). Moreover, $\cos\theta$ decreases as θ increases to π; hence, v will also decrease as θ increases. Note that v is zero (see point B in Figure 7.2.3) when

$$\theta = \arccos(-\frac{a}{2w^2}) < \pi$$

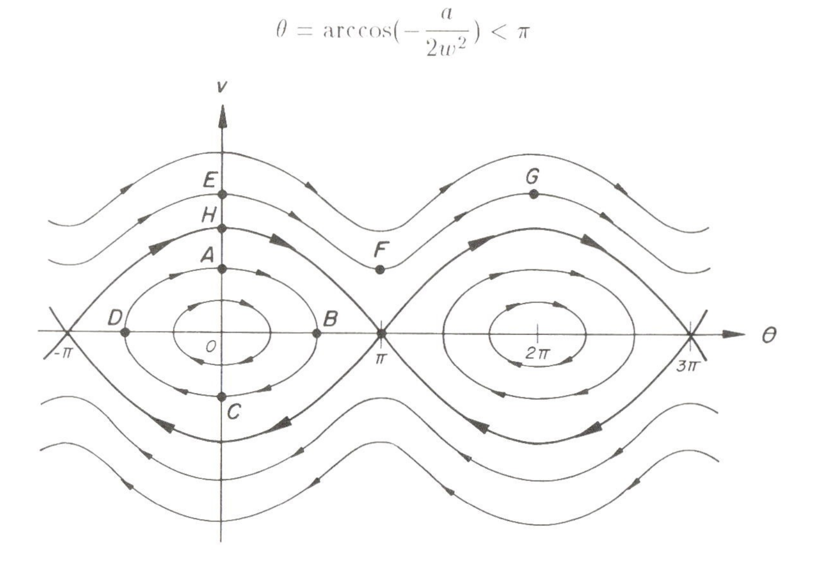

Figure 7.2.3: **Phase plane curves for an undamped simple pendulum**

Using the symmetry of the graph, we obtain a graph like the one marked $ABCD$ in Figure 7.2.3. For different initial values (in other words, different values of a) we shall get different graphs, but they will all have the same ellipse-like shape. When a is near $-2w^2$ then the points A and B will be close to the origin, and as a increases, so will the intercepts A and B on the two axes. Note that the slope is undefined at the points where the graph crosses the θ-axis.

Since the function $\cos\theta$ is periodic with period 2π, the same ellipse-like graphs will also exist around the points $(2n\pi, 0)$ where n is any integer.

Case (iv): $a = 2w^2$

The equation (7.2.10) now has the special form in the first quadrant

$$v = \sqrt{2}w\sqrt{\cos\theta + 1} = 2w\cos\frac{\theta}{2} \qquad (7.2.11)$$

This graph is just part of a cosine function with maximal value $2w$ (see the point H in Figure 7.2.3) and zero at $\theta = \pi$. The rest of the graph follows from symmetry and the periodicity of $\cos\theta$. Note that the slope at the points $(n\pi, 0)$ where n is an odd integer is $\pm w$.

Case (v): $a > 2w^2$

Now v can never be zero in equation (7.2.10), since $|\cos\theta| \leq 1$. The slope of the graph by (7.2.8) is zero at $\theta = n\pi$ where n is a nonnegative integer (see, for example, the points E, F, and G in Figure 7.2.3). The rest of the graph in the other three quadrants follows again from symmetry.

Our analysis of the equation (7.2.9) has shown that there are four different types of curves in the phase plane, depending on the value of a, or equivalently the initial values in (7.2.5). These curves are collectively shown in Figure 7.2.3. The arrows shown on the graphs depict the direction of increasing values of the parameter t: when $v > 0$ it follows by the first equation in (7.2.6) that θ is an increasing function of t, and hence all the arrows in the upper half plane point to the right. By the same argument, all the arrows in the lower half plane point to the left.

Let us leave the analysis of the mathematical model for the moment and turn to the *interpretation stage*. What does Figure 7.2.3 tell us about the physical problem of the motion of a simple pendulum?

Suppose the pendulum is at rest in the position of static equilibrium (the position OA in Figure 7.2.2). In terms of the initial values in (7.2.5) this means that $\alpha = 0$. At time $t = 0$ the pendulum is struck sharply so that it starts to move anticlockwise with velocity β (thus β is positive). Now the constant a in (7.2.9) is

$$a = \beta^2 - 2w^2 \cos\alpha = \beta^2 - 2w^2$$

and so we have

$$\beta = \sqrt{a + 2w^2}.$$

Refering to the analysis above, note that Case (i): $a < -2w^2$ cannot occur here. Case (ii): $a = -2w^2$ corresponds to $\beta = 0$, which means that the pendulum hangs motionless for $t > 0$.

Case (iii): $-2w^2 < a < 2w^2$, when β is positive, is more interesting. When $\beta < 2w$ (see AB in Figure 7.3.2), the graph shows that θ increases and v decreases until $v = 0$. This means that the pendulum swings anticlockwise until it comes momentarily to rest. Then the graph (see BC) shows that θ decreases to zero and v (which is now negative) decreases further to $-\beta$, which means that the pendulum swings in a clockwise direction until it reaches the position of static equilibrium at the initial speed (but in the opposite direction!). Now the graph (see CD) shows that θ becomes negative and $|v|$ diminishes to zero; this implies that the pendulum continues it swing in a clockwise direction until it comes momentarily to rest. Similarly, the graph (see DA) shows that the pendulum returns to the initial position at exactly the same velocity. And as time goes on, this motion is repeated over and over again - perpetual motion, since we assumed no friction or air resistance!

Apart from the motion of the pendulum, we have learned an important thing: *when a graph (or orbit) is closed in the phase plane, then the motion is periodic.*

Suppose that $\beta > 2w$, which is Case (v): $a > -2w^2$ in the analysis above (see also EFG on the graph). Now v is never zero, which means that the pendulum reaches the vertical position ($\theta = \pi$ and point F on the graph) with a positive speed. Hence, the pendulum continues to move anticlockwise for $\pi < \theta < 2\pi$ (see FG on the graph) to arrive at the position of static equilibrium with the same initial velocity and $\theta = 2\pi$, having completed one revolution. Obviously the process now repeats itself as θ gets bigger and bigger.

So we come to the conclusion that an initial velocity less than $2w$ will result in a motion where the pendulum swings to and fro; and an initial velocity bigger than $2w$ will result in the pendulum rotating around the point O.

What happens when $\beta = 2w$? If we look at the graph (starting at point H), the pendulum reaches the vertical position ($\theta = \pi$) at zero speed, and it appears as if the pendulum could either swing back clockwise or continue anticlockwise to complete one revolution. In actual fact neither will happen because the pendulum never really reaches the vertical position! (See Exercise 2 in §7.5.)

We shall not discuss the experimental *validation* of the model, but you can make an interesting experiment at you own risk: Attach a heavy steel ball to a light steel wire of at least 2 meters. Choose a point of suspension so that if the ball is pulled sideways for about 60 degrees, the ball will touch your face between the eyes where you are standing erect with your back to a wall. With the ball just touching your face, let the ball go so that it swings away from you on its arc and then returns. If you do not duck, then you really believe this model! (Warning: If you are a nervous type who might push the ball away from your face so that the ball has a nonzero initial velocity, then you may end up in hospital. The author takes no responsibility for any physical or psychological damage as a result of any mishaps with this experiment!)

Applied mathematicians refer to the graphs in the phase plane in Figure 7.2.3 as *energy curves*. The reason for this is that the expression

$$J(\theta', \theta) = \frac{1}{2}m(\ell\theta')^2 + mg\ell(1 - \cos\theta)$$

is the total energy (the sum of the kinetic and potential energy) of the bob at any time t. If we multiply (7.2.9) by $\frac{1}{2}m\ell^2$, and add the term $mg\ell$, then

$$\begin{aligned} J(\theta', \theta) &= \frac{1}{2}m\ell^2 a + mg\ell \\ &= J(\beta, \alpha) \end{aligned}$$

Hence, (7.2.9) just says that the total energy remains constant on each curve in Figure 7.2.3.

Although we did not find the solution of the boundary value problem (7.2.6), (7.2.5) explicitly as a function of t, we could utilize the fact that the system is autonomous to extract a lot of useful information from the phase plane. We can do even better if we integrate (7.2.9) with respect to t.

Let us, for example, look at Case (iii) where the pendulum oscillates around the position of static equilibrium. The obvious question is what the period τ of this motion is. By symmetry it is clear that we need only calculate the time for the motion in the first quadrant of Figure 7.2.3 (that is, the motion from the origin to the maximal displacement or amplitude θ_m), and multiply the answer by 4. Note also that when $\theta = \theta_m$ we have $v = 0$, and hence, $a = -2w^2 \cos\theta_m$ by (7.2.10). On integrating (7.2.10) we find

$$\left.\begin{aligned} \tau &= 4\int_0^{\theta_m} \frac{d\theta}{\sqrt{2w^2\cos\theta + a}} \\ &= \frac{2\sqrt{2}}{w}\int_0^{\theta_m} \frac{d\theta}{\sqrt{\cos\theta - \cos\theta_m}} \end{aligned}\right\} \qquad (7.2.12)$$

An appropriate numerical procedure (see §3.9) must be used to calculate this integral, which is an example of a so-called *elliptic integral* (see Exercise 4 in §7.5). There are also tables available for these elliptic integrals.

Finally, note that the equation (7.2.8) defines a direction field (see §1.3) at every point in the phase plane, except where $\sin\theta$ and v vanish simultaneously. These exceptional points are called *singular points* or, briefly, *singularities*. In the case of the simple pendulum these singularities are the points $(n\pi, 0)$ with n any integer. The singularities of such an autonomous system differ in their effect on the physical system which it models.

For example, each of the singularities at the points $(2n\pi, 0)$, where n is any integer, is called a *center*, and the other singularities are called *saddles*. Each center is *stable* in the sense that if the motion is initially near the center, then the motion will stay close to the center for $t > 0$. On the other hand, each saddle is *unstable* because if the motion is initially near the saddle, it can move away from the saddle. We can also think of a stable point as a point where slightly different initial values will not result in large differences in the subsequent motion, whilst in the case of an unstable point the motions may differ completely. This agrees with the initial position of a pendulum near the position of static equilibrium or near the vertical position $\theta = \pi$, respectively.

- Read: [9] page 135 or [10] page 377 or [42] page 440 or [50] page 308 or [59] page 14, 24. A nice application of a phase plane is the study of the hammer of an electric bell in [1]. An interesting model of a pendulum with a spring at the top and an exterior force is discussed in [29].
- Do: Exercises 2, 3, 4 in §7.5.

7.3 The Pendulum with Damping

Let us assume that friction and air resistance can be incorporated into the model of §7.2 by adding in (7.2.2) an extra term which is directly proportional to the velocity. Then (7.2.3) will change to

$$\theta''(t) = -c\theta'(t) - w^2 \sin\theta(t) \tag{7.3.1}$$

where c is a positive constant, and (7.2.8) will now read

$$\frac{dv}{d\theta} = \frac{-cv - w^2 \sin\theta}{v} \tag{7.3.2}$$

with the initial values in (7.2.5) unchanged. Unfortunately, this equation is not separable any more and cannot be solved analytically. However, by utilizing the information of §7.2, we can get a qualitative idea of the motion. Note first of all that the set of singular points of (7.2.8) is also the set of singular points of (7.3.2). If we compare the direction field of (7.2.8) with the direction field of (7.3.2), we find that

$$\{\text{ Slope in } (7.3.2)\} = \{\text{ slope in } (7.2.8)\} - c$$

which means that the slope in (7.3.2) is always less than the slope in (7.2.8) at the same point in the phase plane. For example, if we start at the point A of Figure 7.3.2 (the case of $\beta < 2w, \alpha = 0$), the graph of (7.3.2) will cut across the ellipse-like graphs of (7.2.8) and not pass through the points B, C, and D, as shown in Figure 7.3.1.

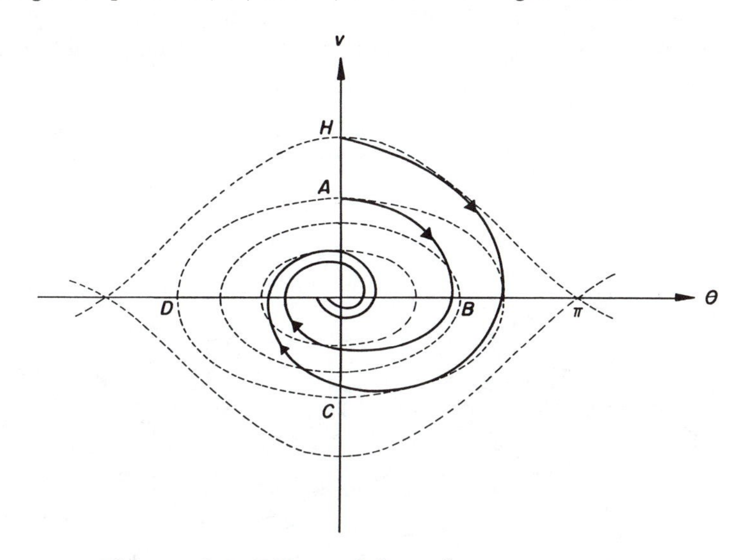

Figure 7.3.1: **Effect of damping on a center**

Hence, the centers at the points $(2n\pi, 0)$ with n any integer will now change to *spirals* which are obviously stable. Note that if we start at the point H in Figure 7.2.3 (the case of $\beta = 2w, \alpha = 0$), the graph will also spiral in to the origin.

As β is increased from $2w$, there must be a value γ at point I on the v-axis so that the graph through I also passes through the point $(\pi, 0)$ (see Figure 7.3.2). When β is increased beyond γ, there must be a value σ at point J on the v-axis so that the graph through J also passes through the point $(3\pi, 0)$. Any value of β between γ and σ (see the point K in Figure 7.3.2) will produce a graph like KLM which spirals in to the point $(2\pi, 0)$. The important difference between (7.2.8) and (7.3.2) is that, with the exception of the graphs passing through the saddle points, all the other graphs spiral in to one of the points $(2n\pi, 0)$ with n an integer. There are no closed graphs and thus no periodic motion. In terms of the pendulum it means that if $\beta < 2w$, the pendulum will swing to and fro with diminishing "amplitude". When β lies between γ and σ, the pendulum completes a full revolution and then swings to and fro. In the case of $\beta = \sigma$, the pendulum completes one revolution before it creeps up on the vertical position.

It is important to know the values γ and σ of the initial velocity because firstly, these initial values lead to instability and secondly, they are the separation between different modes of motion. We could use a numerical algorithm (see Chapter 3) to draw the graphs of (7.3.2) and use a shooting method of different initial values to see if we can strike the point $(\pi, 0)$. We can also use the following technique of a power series, which is frequently used to find numerical solutions of differential equations.

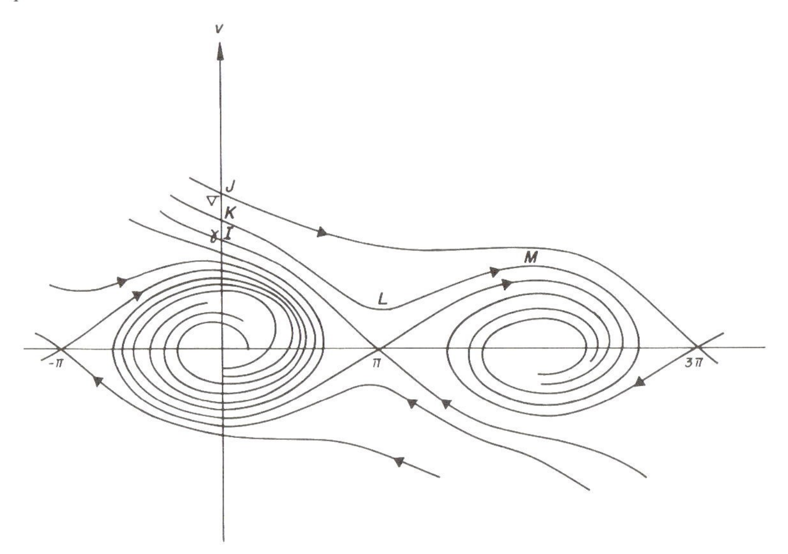

Figure 7.3.2: **Phase plane of a simple pendulum with damping**

Suppose the solution of (7.3.2) is analytic. This means that the solution can be written as a power series which converges over some interval on the real line. Since we must use the condition $v(\pi) = 0$, it is necessary to write the power series in the form

$$v(\theta) = a_0 + a_1(\theta - \pi) + a_2(\theta - \pi)^2 + \ldots\ldots. \qquad (7.3.3)$$

where the coefficients a_i are as yet undetermined constants. It is, however, immediately clear that $a_0 = 0$. It is also convenient to introduce a new independent variable

$$x = \theta - \pi \qquad (7.3.4)$$

to simplify the calculations. Now $v(-\pi) = \gamma$. The transformed equation (7.3.2) can then be written in the form

$$v(\frac{dv}{dx} + c) = w^2 \sin x. \qquad (7.3.5)$$

Note that the right hand side is an odd function. Hence, we can take $v(x)$ also as an odd function, since the expression in brackets on the left hand side will then be an even function. Thus, the series in (7.3.3) can be written

$$v(x) = a_1 x + a_3 x^3 + a_5 x^5 + ... \quad (7.3.6)$$

Substitute (7.3.6) in (7.3.5):

$$(a_1 x + a_3 x^3 + ...)(c + a_1 + 3a_3 x^2 + ...) = w^2 (x - \frac{x^3}{3!} + ...) \quad (7.3.7)$$

where the Maclaurin series for $\sin x$ was used (see §2.14).

By equating coefficients of like powers of x, we now obtain a set of equations from which the constants a_i can be calculated. Starting with the coefficients of x we find that

$$\left.\begin{array}{rcl} a_1^2 + ca_1 - w^2 & = & 0 \\ a_1 & = & \dfrac{-c \pm \sqrt{c^2 + 4w^2}}{2} \end{array}\right\} \quad (7.3.8)$$

which gives one positive and one negative value for a_1. Note in Figure 7.3.2 that there are two solutions passing through the point $(\pi, 0)$, one with a positive slope and one with a negative slope. If we differentiate (7.3.6) to find the slope at $x = 0$ (or equivalently the slope at $\theta = \pi$), we see that a_1 is precisely this slope, and it is now clear why there are two values for a_1 in (7.3.8). Since we want to calculate γ, we must select the *negative* value of a_1 in (7.3.8). Next we look at the coefficients of x^3 in (7.3.7):

$$\begin{array}{rcl} 3a_1 a_3 + (c + a_1)a_3 & = & \dfrac{w^2}{3!} \\ a_3 & = & -\dfrac{w^2}{6(c + 4a_1)} \end{array}$$

Similarly, we find that

$$a_5 = \frac{w^2 - 360a_3^2}{120(c + 6a_1)}$$

Suppose that the damping constant amounts to $c = 0.2w$, where we express the damping constant in terms of w on purpose to compare our result with the result of $\beta = 2w$ (the point H in Figure 7.2.3) of §7.2. For this value of c the power series (7.3.6) is

$$v(x) = -1.1050wx + 0.0395wx^3 - 0.0006wx^5 - .... \quad (7.3.9)$$

When the first three terms in this series are used, we find that

$$\gamma = v(-\pi) \approx 2.4w$$

The accuracy of this answer is left to the exercises. Note that the curve passing through $\theta = \pi$ in Figure 7.3.2 is approximated by the polynomial of the fifth degree

$$v(\theta) = -1.1050w(\theta - \pi) + 0.0395w(\theta - \pi)^3 - 0.0006w(\theta - \pi)^5$$

In the neighbourhood of the point $(\pi, 0)$ the approximation will be very good, but as θ moves away from π, the two curves will separate. Hence, if the approximation to γ is acceptable, then the approximation at any $\theta \in [0, 2\pi]$ is also acceptable or better. By taking more terms in (7.3.9) we could, of course, improve the approximation.

By comparing the differential equation (7.3.2) with the equation (7.2.8), we could utilize our knowledge of the latter equation to obtain qualitative information about (7.3.2). It was also demonstrated that power series can be used to approximate the solutions of (7.3.2).

- Read: [10] page 396 or [42] page 452 or [50] page 321 or [59] page 62.
- Do: Exercises 5, 6 in §7.5.

7.4 Population of Interacting Species

The first model discussed in Chapter 2 was a simple model of population dynamics. We shall end this book by considering the more complicated model of two different populations coexisting within the same living space, but one population influencing the birth/death of the other population. There is extensive literature on the growth and decline of two interacting populations, from both a biological and a mathematical viewpoint, mainly because of the obvious importance of these studies for the ecology.

We shall restrict ourselves to the case where one population can only grow by killing the other population. These models are referred to as *predator-prey models*. In general, a cycle develops: if the prey is numerous, then the population of the predators will increase to the point where the stock of prey cannot support all the predators, with the result that the number of predators decreases, which, in turn, allows the prey to increase again (very much like the love/hate relationship between

William and Zelda in §5.3). Let us look at a two different situations which both lead to the same mathematical model.

Reproduction

The wasp *Hemipepsis capensis* attacks large groundspiders fearlessly, paralyses them by stinging, buries them, and then lays an egg on them. The larva from the egg lives on the spider through the winter, before changing into a pupa in the summer, with the adult wasp emerging 2 or 3 weeks later. This interaction can be modelled as follows (see [62] page 447):

Let $x(t)$ and $y(t)$ denote the number of spiders and wasps, respectively, alive at time t in the ecosystem under consideration. Define also

$q =$ the annual birth percentage of the spiders;
$s =$ the annual natural mortality percentage the spiders;
$r =$ the annual natural mortality percentage of the wasps.

The following assumptions are necessary for the construction of the model:

Assumptions (W)

(1) *x and y are continuously differentiable functions of t for $0 < t < \infty$;*
(2) *$q, s,$ and r are constants;*
(3) *the number of eggs deposited in spiders at time t is proportional to the number of encounters between wasps and spiders at that time;*
(4) *the probablity that a spider and a wasp will meet at any time t is constant;*
(5) *there is a fixed probability h that the death of a spider due to the laying of an egg will result in the birth of a wasp;*
(6) *the time lag between the death of a spider and the consequent birth of a wasp is zero;*
(7) *neither spiders nor wasps enter or leave the ecosystem for $0 < t < \infty$.*

The justification for Assumption (W1) is that we are dealing with large populations (see §2.2 and especially Figure 2.2.1). As in §2.7, we argue that the number of encounters between x spiders and y wasps in the time interval $[t, t+\delta t]$ is proportional to the product $xy\delta t$. Hence, by Assumptions (W3) and (W4) the number of deaths of spiders in the time interval $[t, t+\delta t]$ is $kxy\delta t$. Also by Assumptions (W5) and (W6) the number of births of wasps in the time interval $[t, t+\delta t]$ is $hkxy\delta t$.

We can now equate the change in the populations of the spiders and the wasps in the time interval $[t, t+\delta t]$ to the difference between their births and deaths (using Assumption (W7)).

$$\left.\begin{aligned} x(t+\delta t) - x(t) &= \frac{qx}{100}\delta t - \frac{sx}{100}\delta t - kxy\delta t \\ y(t+\delta t) - y(t) &= hkxy\delta t - \frac{ry}{100}\delta t \end{aligned}\right\} \tag{7.4.1}$$

For convenience, let us write

$$g = \frac{q-s}{100} \qquad \text{and} \qquad p = \frac{r}{100} \tag{7.4.2}$$

Here the fraction g multiplied by the number of spiders is their net natural annual increase, and the fraction p multiplied by the number of wasps is their natural annual mortality. Divide (7.4.1) by δt and let $\delta t \to \infty$, then we get

$$\left.\begin{aligned} \frac{dx}{dt} &= gx - kxy \\ \frac{dy}{dt} &= hkxy - py \\ x(0) &= \alpha, \qquad y(0) = \beta \end{aligned}\right\} \tag{7.4.3}$$

with α and β the initial population of the spiders and the wasps, respectively. Note that k, h, p, α, and β are all positive constants.

Food

In the food chain in nature, predators have to hunt for their food or die of starvation. This is true of spiders hunting insects, lions hunting deer, or sharks feeding off fish. We shall consider a Malthus model (see §2.2) for the prey in the absence of predators, as we had done for the spiders in the spider/wasp model above.

Let $x(t)$ and $y(t)$ denote the number of the prey and predators, respectively, alive at time t in the ecosystem under consideration. Define also

q = the annual birth percentage of the prey;

s = the annual natural mortality percentage of the prey;

r = the annual natural mortality percentage of the predators.

The following assumptions are necessary for the construction of the model:

Assumptions (X)

(1) *x and y are continuously differentiable functions of t for $0 < t < \infty$;*
(2) *q, s, and r are constants;*
(3) *the percentage birth rate of the predators is proportional to the number of prey available at time t;*
(4) *the percentage mortality rate of the prey due to the predators is proportional to the number of predators active at time t;*
(5) *the time lag between the death of one of the prey and the consequent birth of one of the predators is zero;*
(6) *neither prey nor predators enter or leave the ecosystem for $0 < t < \infty$.*

As in the spider/wasp model above, using (7.4.2) and denoting the proportionality constants in Assumptions (X3) and (X4) by hk and k, respectively, we find that the change in the populations in the time interval $t, t + \delta t]$ is

$$\begin{aligned} x(t+\delta t) - x(t) &= gx\delta t - kxy\delta t \\ y(t+\delta t) - y(t) &= hkxy\delta t - p\delta t \end{aligned}$$

which again gives the initial value problem (7.4.3) where α and β are now the initial population of the prey and predators, respectively.

Analysis of the Model

The system (7.4.3) is nonlinear. We cannot express the solution pair $(x(t), y(t))$ as functions of t. However, we can again use the phase plane to obtain important qualitative information about the solution. By (7.4.3)

$$\frac{dy}{dx} = \frac{hkxy - py}{gx - kxy} \tag{7.4.4}$$

This equation is separable (see §2.1) and hence, for $x > 0, y > 0$

$$\int_\beta^y \frac{g - k\eta}{\eta} d\eta = \int_\alpha^x \frac{hk\xi - p}{\xi} d\xi$$

$$g\ln y - g\ln\beta - ky + k\beta = hkx - hk\alpha - p\ln\xi + p\ln\alpha.$$

For convenience define the constant

$$c = g\ln\beta - k\beta - hk\alpha + p\ln\alpha$$

where c is determined by the initial conditions α and β, to obtain the solution of (7.4.4):

$$g\ln y + p\ln x - ky - hkx = c \tag{7.4.5}$$

To plot this graph is hard work. For every given value of x, a numerical procedure must be used to find the corresponding values of y. Let

$$c_0 = g\ln\frac{g}{ke} + p\ln\frac{p}{hke} \tag{7.4.6}$$

We show in §7.6 that

* If $c > c_0$, then there are no real values of x and y which satisfy (7.4.5).

* If $c = c_0$, then the graph of (7.4.5) is a single point $(\frac{p}{hk}, \frac{g}{k})$.

* If $c < c_0$, then there are exactly two values of y for every x in an open interval $I(c)$, and one value $y = \frac{g}{k}$ for x at the endpoints of $I(c)$.

An example of the graphs of (7.4.5) for different values of c such that $c \leq c_0$ is shown in Figure 7.4.1. Note that for each value of c, the minimum and maximum of x is at $y = \frac{g}{k}$, and the minimum and maximum of y is at $x = \frac{p}{hk}$.

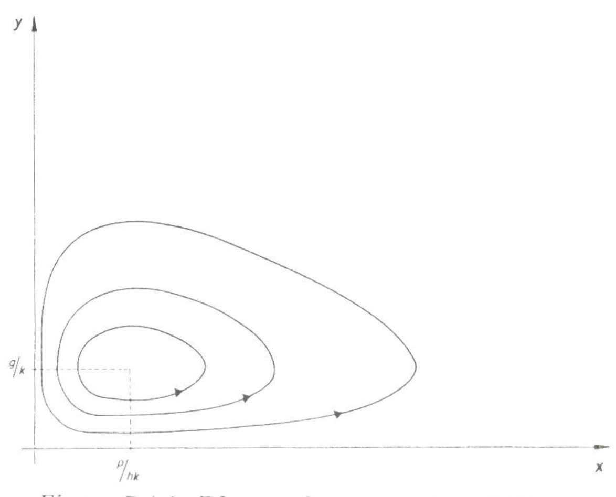

Figure 7.4.1: **Phase plane graphs of (7.4.3)**

There is only one singular point $(\frac{p}{hk}, \frac{g}{k})$ where the slope of (7.4.4) is undefined. (We had assumed that x and y are positive in the above analysis, and anyhow we are not interested in the origin which implies zero populations.) At this point we have $x'(t) = 0$ and $y'(t) = 0$, which means that there is no increase or decrease in either of the two populations. Hence, we call these values

$$\left.\begin{aligned} a &= \frac{p}{hk} \\ b &= \frac{g}{k} \end{aligned}\right\} \qquad (7.4.7)$$

the equilibrium population of each species. The closed curves around this singular point (a, b) in Figure 7.4.1 show that neither population tends to its equilibrium population, but fluctuates periodically around it.

We can do some more analysis for the case when the initial values α and β are chosen near the equilibrium populations a and b, respectively. Now the deviations $x - a$ and $y - b$ are small, and this enables us to make some approximations. Let us introduce new variables

$$\left.\begin{aligned} X &= \frac{x-a}{a} \\ Y &= \frac{y-b}{b} \end{aligned}\right\} \qquad (7.4.8)$$

which means that we shift the x and y axes parallel to themselves to a new origin at the singular point (a, b), and we change the scale on the axes. We can think of X and Y as normalized deviations.

Substituting (7.4.8) in (7.4.3), we obtain the following initial value problem:

$$\left.\begin{aligned} \frac{dX}{dt} &= -gY - gXY \\ \frac{dY}{dt} &= pX + pXY \\ X(0) &= \frac{\alpha - a}{a} \\ Y(0) &= \frac{\beta - b}{b} \end{aligned}\right\} \qquad (7.4.9)$$

By the choice of α and β, these initial values $X(0)$ and $Y(0)$ are small. The closed curves in Figure 7.4.1 show that $X(t)$ and $Y(t)$ are then also small for $t > 0$. Since the product XY is much smaller than either X

or Y, we can approximate (7.4.9) by the linear system

$$\left.\begin{aligned}\frac{dX}{dt} &= -gY \\ \frac{dY}{dt} &= pX\end{aligned}\right\} \qquad (7.4.10)$$

(This should look familiar to you - remember William and Zelda in §5.3?) In the phase plane we now have

$$\frac{dY}{dX} = -\frac{p}{g}\frac{X}{Y}$$

which is separable, and after integration

$$\left.\begin{aligned}gY^2 + pX^2 &= C \\ &= g\left(\frac{\beta - b}{b}\right)^2 + p\left(\frac{\alpha - a}{a}\right)^2.\end{aligned}\right\} \qquad (7.4.11)$$

These curves are ellipses which then approximate the orbits of Figure 7.4.1 in the neighborhood of the singular point.

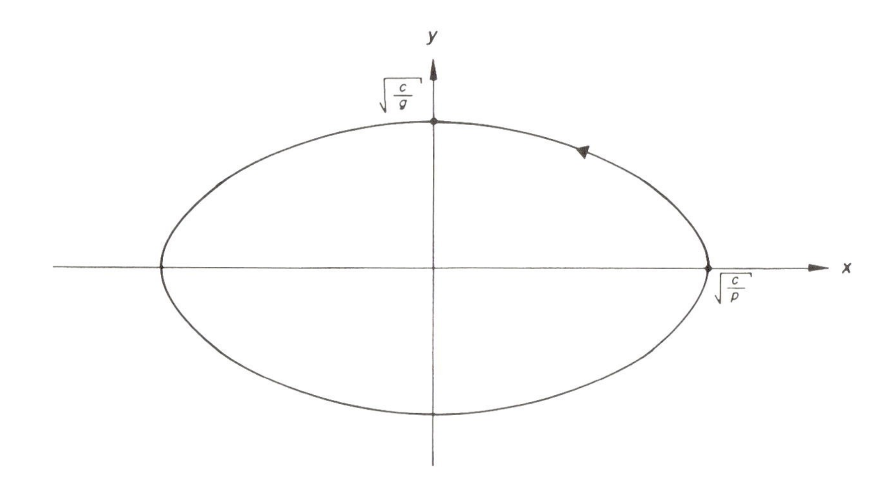

Figure 7.4.2: **Approximation near the singular point**

As before, we can interpret the analysis in terms of the two populations. When the deviation X is a maximum of $\sqrt{\frac{C}{p}}$, the deviation of Y is zero. Hence, there are many spiders/prey available at a time when the wasps/predators are at their equilibrium population. Moving anticlockwise on the ellipse, the population of spiders/prey decreases, and the population of wasps/predators increases, until the deviation Y reaches a maximum of $\sqrt{\frac{C}{g}}$. Now (in the fourth quadrant) spiders/prey are getting scarce, so that the population of wasps/predators starts to decline. When the wasps/predators reach their equilibrium population, the spiders/prey are at a minimum, but since the numbers of wasps/predators keep on going down, the spiders/prey start to increase again (in the third quadrant). After the wasps/predators have reached their minimum, they start to increase again (in the second quadrant) until the starting point is reached and the whole process repeats itself.

We can, of course, also solve (7.4.10) directly as a function of t:

$$X'' = -gY' = -pgX(t)$$

which is a simple harmonic motion (see §6.1)

$$X(t) = A\cos(\sqrt{pg}t + \phi).$$

Substitute this in the first equation of (7.4.10) to obtain

$$Y(t) = \sqrt{\frac{p}{g}}A\sin(\sqrt{pg}t + \phi)$$

The constants A and ϕ follow from the initial conditions

$$\begin{aligned} A\cos\phi &= \frac{\alpha - a}{a} \\ A\sin\phi &= \sqrt{\frac{g}{p}}\frac{\beta - b}{b} \end{aligned}$$

Provided that α and β are near the equilibrium populations, we have thus shown that the populations can be approximated by

$$\left.\begin{aligned} x(t) &= \frac{p}{hk} + \frac{pA}{hk}\cos(\sqrt{pg}t + \phi) \\ y(t) &= \frac{g}{k} + \frac{\sqrt{pg}A}{k}\sin(\sqrt{pg}t + \phi) \end{aligned}\right\} \qquad (7.4.12)$$

We can interpret this result as follows:

Firstly, both populations vary with period

$$T = \frac{2\pi}{\sqrt{pg}} = \frac{200\pi}{\sqrt{r(q-s)}}$$

which depends on the percentage birth and mortality rates of the species, but *not* on the initial populations.

Secondly, the wasp/predator population trails the spider/prey population by a quarter cycle (or $\frac{\pi}{2\sqrt{pg}}$ radians out of phase). Thirdly, since

$$A = \sqrt{(\frac{\alpha}{a} - 1)^2 + \frac{g}{p}(\frac{\beta}{b} - 1)^2}$$

is a function of the initial values, then also so are the amplitudes for both populations. The initial value problem (7.4.3) is known in the literature as the Lotka-Volterra model. By changing the Assumptions (W) or (X) more realistic models can be constructed, but the analysis becomes more complex.

- Read: [49] page 66 or [28] page 224 about different models. There are many interesting papers on interacting populations. A vicious circle in nature is modelled with linear equations in [52], and a study of deer and wolves on an island can be found in [30]. The interaction of two types of fruitfly is also discussed in [4].

7.5 Exercises

(1) Write the initial value problem

$$\begin{aligned} \frac{d^2x}{dt^2} + \sec(\frac{dx}{dt}) &= 0 \qquad (0 < t < 1) \\ x(0) = x'(0) &= 0 \end{aligned}$$

as a system of first order equations.

(a) Use the system to find a solution $x(t)$.

(b) Discuss the uniqueness of this solution.

(c) Does a solution exist for $t > 1$?

(2) (a) Integrate (7.2.11) to show that

$$t = \frac{1}{w} \ln(\sec\frac{\theta}{2} + \tan\frac{\theta}{2})$$

(b) Show that $t \to \infty$ as $\theta \to \pi$.

(c) Show that the equation in (a) can be written as

$$\theta = 4 \arctan(\frac{e^{wt} - 1}{e^{wt} + 1}).$$

(Hint: use the half angle formula for each of the sine and cosine functions.)

(3) Suppose a simple pendulum is initially at $\theta = 45°$ with a velocity of β. The length of the pendulum is 40 centimeters. Use the model of §7.2 to calculate β so that the amplitude of the resulting motion would be (i) 90° (ii) 125°.

(4) Introduce a new variable z in the integral (7.2.12) by means of the transformation

$$\sin\frac{\theta}{2} = \sin\frac{\theta_m}{2}\sin z.$$

(a) Show that the integral can be written as

$$\tau = \frac{4}{w}\int_0^{\frac{\pi}{2}} \frac{dz}{\sqrt{1 - sin^2\frac{\theta_m}{2}\sin^2 z}}$$

which is known as a *complete elliptic integral of the first kind.*

(b) Use the Maclaurin series (see §2.14) for $(1 - x^2)^{-\frac{1}{2}}$ to show that

$$\tau = \frac{2\pi}{w}[1 + \frac{1}{2^2}\sin^2\frac{\theta_m}{2} + (\frac{1 \times 3}{2 \times 4})^2 \sin^4\frac{\theta_m}{2}..........]$$

(c) Write a program to calculate θ_m and τ if the length of the pendulum ℓ and the initial values α and β are given.

(5) (a) Find also the coefficients a_7 and a_9 in the power series of (7.3.6) in terms of c and w.

(b) Calculate a_7 and a_9 if $c = 0.2w$.

(c) Is the answer of $\gamma \approx 2.4w$ in §7.3 accurate to one decimal place?

(d) Use the program OURGRAPH or any similar program of your choice to draw the polynomial $\bar{v}(\theta)$ in (7.3.9) on the interval $[0, 2\pi]$ for the two cases of a fifth degree and a ninth degree polynomial.

(6) As in §7.3, use the method of power series to calculate σ in terms of w when $c = 0.2w$.

7.6 Mathematical Background

The following equation described the orbits in the phase plane of (7.4.3):

$$g \ln y + p \ln x - ky - hkx = c \tag{7.6.1}$$

We now show that the following statement about the orbits, made in §7.4, is valid:

* If $c > c_0$, then there are no real values of x and y which satisfy (7.6.1).

* If $c = c_0$, then the graph of (7.6.1) is a single point $(\frac{p}{hk}, \frac{g}{k})$.

* If $c < c_0$, then there are exactly two values of y for every x in an open interval $I(c)$, and one value $y = \frac{g}{k}$ for x at the endpoints of $I(c)$.

The constant c_0 was defined as

$$c_0 = g \ln \frac{g}{ke} + p \ln \frac{p}{hke} \tag{7.6.2}$$

Suppose a value $x = \xi$ is given, and the corresponding value of y in (7.6.1) must be determined. Then

$$\left.\begin{aligned} g \ln y &= ky + c + hk\xi - p \ln \xi \\ &= ky + c_1 \end{aligned}\right\} \tag{7.6.3}$$

The left hand side is the natural logarithm, and the right hand side is a straight line with slope k and c_1 the intercept on the vertical axis, as shown in Figure 7.6.1. It is clear that the tangent when $c_1 = \bar{c}_1$ will have only one point of intersection with the graph of the logarithm, namely $y = \frac{g}{k}$. By (7.6.3) it then follows that

$$\bar{c}_1 = g \ln \frac{g}{k} - g = g \ln \frac{g}{ke} \tag{7.6.4}$$

When $c_1 > \bar{c}_1$ there is no real value of y which will satisfy (7.6.3). On the other hand, when $c_1 < \bar{c}_1$ there are two positive real values of y which will satisfy (7.6.3).

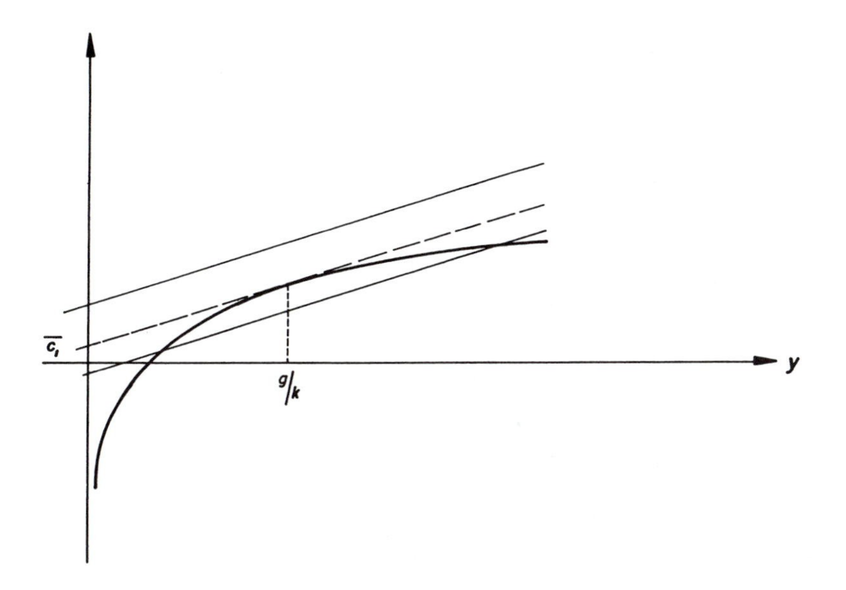

Figure 7.6.1: **The intersection of $\ln y$ with a straight line**

Suppose that $c_1 \leq \bar{c_1}$, then by (7.6.3) and (7.6.4)

$$hk\xi - p\ln\xi \leq g\ln\frac{g}{ke} - c \tag{7.6.5}$$

Let us first look at the graph of

$$z = hkx - p\ln x$$

which is shown in Figure 7.6.2. By differentiation we see that the slope is zero at the point $(\frac{p}{hk}, p - p\ln\frac{p}{hk})$ and changes its sign there. Hence, a real value of ξ will only exist if the right hand side in (7.6.5) is not less than $p - p\ln\frac{p}{hk}$. Thus we have that

$$c \leq g\ln\frac{g}{ke} + p\ln\frac{p}{hke} = c_0$$

When the equality holds, we obtain by the graph only one value $\xi = \frac{p}{hk}$, which in turn will result in an equality in (7.6.5) and thus $c_1 = \bar{c}_1$ in (7.6.3), so that we obtain only one value of y, namely $y = \frac{g}{k}$. When the inequality holds and this value of c is substituted in (7.6.5), we obtain two positive values of ξ when the equality is taken in (7.6.5). These two positive values of ξ are the endpoints of an interval $I(c)$. Due to the equality in (7.6.5), only one value of y then follows by (7.6.3) since $c_1 = \bar{c}_1$ in this case. For any ξ in the open interval $I(c)$, we obtain two values of y by (7.6.3).

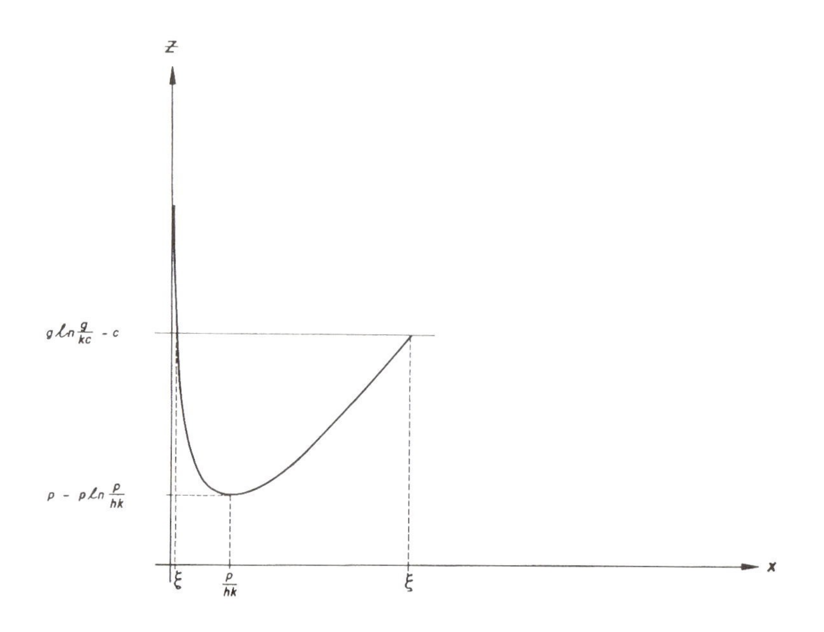

Figure 7.6.2: **The interval** $I(c)$

8

Table of Integrals

(1) $\int \frac{dx}{(ax+b)(cx+d)} = \frac{1}{ad-bc} \ln |\frac{ax+b}{cx+d}| \quad (ad \neq bc)$[2ex]

(2) $\int \frac{dx}{(ax+b)^2} = -\frac{1}{a(ax+b)} \quad (a \neq 0)$

(3) $\int \frac{xdx}{(ax+b)(cx+d)} = \frac{1}{ad-bc}[-\frac{b}{a} \ln |ax+b| + \frac{d}{c} \ln |cx+d|] \quad (ad \neq bc)$

(4) $\int \frac{xdx}{(ax+b)^2} = \frac{1}{a^2}[\ln |ax+b| + \frac{b}{ax+b}] \quad (a \neq 0)$

(5) $\int \frac{dx}{ax^2+bx+c} = \frac{2}{\sqrt{4ac-b^2}} \arctan \frac{2ax+b}{\sqrt{4ac-b^2}} \quad (b^2 < 4ac)$

(6) $\int \frac{xdx}{ax^2+bx+c} = \frac{1}{2a} \ln |ax^2+bx+c| - \frac{b}{a\sqrt{4ac-b^2}} \arctan \frac{2ax+b}{\sqrt{4ac-b^2}}$ $(b^2 < 4ac)$

(7) $\int e^{ax} \cos bx dx = \frac{e^{ax}}{a^2+b^2}(a \cos bx + b \sin bx)$

(8) $\int e^{ax} \sin bx dx = \frac{e^{ax}}{a^2+b^2}(a \sin bx - b \cos bx)$

(9) $\int \arcsin ax dx = x \arcsin ax + \frac{1}{a}\sqrt{1-a^2x^2} \quad (a \neq 0)$

(10) $\int \arccos ax dx = x \arccos ax - \frac{1}{a}\sqrt{1-a^2x^2} \quad (a \neq 0)$

(11) $\int \arctan ax dx = x \arctan ax - \frac{1}{2a} \ln[1+a^2x^2] \quad (a \neq 0)$

(12) $\int x^n \ln ax dx = x^{n+1}[\frac{\ln ax}{n+1} - \frac{1}{(n+1)^2}] \quad (n \neq -1)$

(13) $\int \frac{\ln ax}{x} dx = \frac{1}{2}(\ln ax)^2$

9

Answers

Section 1.5

1.(a) $X(t) = X(0) + kt$

(b) $X(t) \to \infty$ as $t \to \infty$, but the limited amount of space and food in the bottle contradicts this. Hence, the model is not valid for large values of t.

(c) To describe the best fit the vertical distance between the line and each data point must all be positive, otherwise errors cancel out. Use absolute values or squares of the differences. The best fit (for squares) is

$$X(t) = \frac{530}{3} + 102t$$

(and once you have read §2.2 you will know how I know this!) At any rate, the fit is not satisfactory, since the data points clearly show that the graph of $X(t)$ should bend downwards.

(d) $X(t) = (1 + a)^t X(0$

Interpretation: Again $X(t) \to \infty$ as $t \to \infty$, since $a > 0$, so that the model is not valid for large values of t.
Validation: The best fit (for squares) is

$$X(t) = 209(1.288)^t$$

which is unsatisfactory, since $X(5) = 740$.

(e) Go, man, go!

3.(a) (i) 4th order (ii) $-\infty < x < \infty$ (iii) linear

(b) (i) first order (ii) $0 < x < \infty$ (iii) nonlinear

4.(a) No, $y(\pi) = 1$ is not satisfied.

(b) No, $y = |x|$ is not differentiable at $x = 0$

(c) Yes

(d) Yes

7.(a) 0.603

(b) 0.357

(c) 0.655

Section 2.12

1.(a) 146,410

(b) 18.2 hours

2. 50,238

3. 104

4.(a) 0.25 and 3.95

(b) 22.7

5.(a) 0.20 and 4.98

(b) 20.2

6.(a) Yes

(b) No

(c) Yes

(d) Yes

9.(a) 371

(b) 376

10.(a) $\alpha(b-N)^{1+c} = N(b-\alpha)^{1+c}e^{-at}$

(b) $N \to b$

(c) Inflection point at

$$\tau = \frac{1+c}{a}\ln\frac{c(b-\alpha)}{b(c-1+\sqrt{1+c})} + \frac{1}{a}\ln\frac{b(\sqrt{1+c}-1)}{\alpha c}.$$

When $\alpha = b$, then $N(t) = b$ for $t \geq 0$.

11. $m(t) = (\frac{\alpha}{\beta})^3(1 - e^{-\frac{\beta t}{3}})^3$

12. 29.8 years

13.(a) 0.0494 and 1.44×10^{-8}

(b) 1,459,000 metric tons

(c) 3,435,000 metric tons

14.(a) 323

(c) 163 years

15.(a) 311

(b) 141

16. $w(t) = \frac{1-k}{k}(\alpha - \frac{b}{a})(e^{kat} - 1)$ for $0 < k \le 1$;
$w(t) = (a\alpha - b)t$ for $k = 0$;
Assumption: $b \le a\alpha$

17. $k = 0$, \$630,000

18.(a) $(a+1)^{-1}$

(b) $a > 1$

19. 16 days

20. $x(t) = 1 - (1 - \alpha)e^{-kq\alpha - k(1-\alpha)(t-2p) + (1-\alpha^{-1})(e^{-k\alpha(t-2p)} - 1)}$

22. 675

23. β increases

24. 403 meters, 209 km/h

25. 92.45 seconds

27. $mpc^{-2}(\ln 3 - \frac{2}{3})$

28. 65.6 km/h, 51.3 seconds

29. 62.9 km/h

30.(b) $v(t) = 20000 + 10(250 - t) - 20147.62(250 - t)^{0.02}$

(c) $v(200) = -1286.75$ No lift-off!

31.(a) 10,574 meters

(b) It is valid.

32. $y = x^2 + 1$

33.(a) $y = \pm x\sqrt{x^2 + 2}$.

(c) $y = \pm\sqrt{(x^2 - 1)(x^2 + 3)}$; graph not defined for $|x| < 1$.

34.(a) (i) $y = \sqrt{2(\cos x + 1)}$
(ii) $y = \sqrt{2\cos x - 1}$ for $0 \le x \le \frac{\pi}{3}$
(iii) $y = \sqrt{2\cos x + 7}$ for $0 \le x \le \frac{2\pi}{3}$

35.(a) $y = \frac{1}{2}(1 + e^{-x^2})$

(b) $y = 1 + x^2 + \sqrt{1 + x^2}$

36.(a) $x^3 + y^3 = 2xy$

(b) $y^2 + 2xy - 3x^2 = 5$

37. $\ln(v^2 + cvx + kx^2) - \frac{2c}{\sqrt{4k - c^2}} \arctan \frac{2v + cx}{x\sqrt{4k - c^2}} = \text{constant}$

Section 2.13

Project F:

(a) $p(t) = \beta + (\alpha - \beta)e^{-k(b+d)t}$

(b) $p(t) = \alpha e^{-Bt} + \frac{A}{B}(1 - e^{-Bt}) + \frac{E}{w^2 + B^2}(B \sin wt - w \cos wt + we^{-Bt})$
with $A = k(a - c)$, $B = k(b + d)$ and $E = kf$.

(c) $p(t) = \frac{c-a}{b+d}(e^{\frac{b+d}{m-n}t} - 1) + \alpha e^{\frac{b+d}{m-n}t}$

Section 3.7

1. $y = \frac{t^2}{4}$ and $y = 0$ for $t \ge 0$. The function $f(t, y) = \sqrt{y}$ is not Lipschitz on any interval containing the point $y = 0$.

2. $y = \frac{2}{2 - t^2}$, $|t| < \sqrt{2}$

3. $\delta = 0.5$, $0 \le y \le 2$

4. The five values for the approximation are:
$1, 1, 1.01, 1.0304, 1.0623$.

5. The nine values for the approximation are:
$1, 1, 1.0025, 1.0075, 1.0151, 1.0254, 1.0386, 1.0548, 1.0742$.

6. $y = t + e^{-t}$

10. $a(b - c)$

12. $6.129, \quad y = \frac{1}{1-t}$

13.(a) $3.94, 3.27, 2.81$

(c) $|error| \leq 4.06h, \quad h \leq 0.00025$

14.(a) $0.48, 1.16, 1.71, 2.18, 2.61$

(c) $|error| \leq 1.13$

19. $|e_i| \leq 1,987,975h^4$

Section 4.5

2. $(s - k)^{-1} \qquad (s > k)$

3. $2s^{-3} \qquad (s > 0)$

4. $k(s^2 - k^2)^{-1} \qquad (s > |k|)$

5. $\sqrt{\pi}s^{-\frac{1}{2}} \qquad (s > 0)$

6.(a) Yes

(b) No; undefined at $t = 0$.

(c) Yes

(d) No; undefined at $t = 2$.

(e) Yes

10.(a) Yes

(b) No; not defined at $t = 1$.

(c) Yes

(d) Yes

(e) No; undefined at $t = 0$.

14. $2(s - k)^3$

16.(a) t

(b) $2t - 2\sin t$

17.(a) $a^{-2}(e^{at} - 1) - \frac{t}{a}$

(b) $e^{4t} - (3t+1)e^{t}$

18.(a) $2t + 4e^{-t}$

(b) $0.011e^{7t} + 1.889e^{-2t} - 1.900e^{-3t}$

(c) $4\cos t + 3\sin t - e^{t}(\cos 2t + 0.5\sin 2t)$

(d) $2e^{-t} + e^{-2t}(\sin 3t - \cos 3t)$

19.(a) te^{1-t}

(b) $t + \pi\cos t + k\sin t$ (k an arbitrary constant)

(c) $t + \pi\cos t$ (d) no solution.

20. $0.045(e^{t}\cos 2t - 1) + 0.477e^{t}\sin 2t$

21.(a) $0.5t^2$

(b) kt^2e^{-t} (k an arbitrary constant)

22. $\cos t$

23.(a) $2te^{-t}$

(b) $\sin 2t$

(c) $t - 0.5t^2$

25. 3.323

Section 5.8

1. Only one initial value is given.

4. Trajectory: $y = -\frac{g}{2V^2}(x\sec\theta)^2 + x\tan\theta$;
Height $= \frac{(V\sin\theta)^2}{2g}$.

5. (i) 19.71° (ii) 13.91° (iii) 43.37°

7. $z(t) = \beta\cosh(\sqrt{ab}t) + \frac{b\alpha}{\sqrt{ab}}\sinh(\sqrt{ab}t)$;
$w(t) = \alpha\cosh(\sqrt{ab}t) + \frac{a\beta}{\sqrt{ab}}\sinh(\sqrt{ab}t)$; No.

8. $y(x) = \frac{\alpha \sin p(a-x)}{\cos pa}$; $z(x) = \frac{\alpha \cos p(a-x)}{\cos pa}$;
Critical length $= \frac{\pi}{2p}$.

9.(a) Let $r = Q^{-1} - q^{-1}$.

$$\begin{aligned} U(x) &= \alpha \frac{Qe^{-kra} - qe^{-krx}}{Qe^{-kra} - q} \\ u(x) &= Q\alpha \frac{e^{-kra} - e^{-krx}}{Qe^{-kra} - q} \end{aligned}$$

(b) a as long as possible; q as high as possible.

10. $i(t) = 10(1 - e^{-5t})$.

11. $q(t) = 0.25(1 - e^{-8t})$ coulomb; $i(t) = 2e^{-8t}$ ampère.

12. $q(t) = 10 - 5(1 + 0.02t)^{-500}$ coulomb;
$i(t) = 50(1 + 0.02t)^{-501}$ ampère for $0 \le t \le 1000$.

13. Current through switch $= 25 - 25e^{-4t} + \frac{3}{2}e^{-t}$;
current through inductor $= 25 - 25e^{-4t}$;
charge on capacitor $= \frac{1}{2}(5 - 3e^{-t})$.

14. $i_1(t) = \frac{1}{15}(-11e^{-5t} - 22e^{-20t} + 33\cos 10t + 33\sin 10t)$;
$i_2(t) = \frac{1}{3}(22e^{-5t} - 22e^{-20t} - 33\sin 10t)$.

15.(a) $q_3(t) = 0.581\sin 50t - 0.105\cos 50t - 0.345e^{-20t} + 0.449e^{-80t}$;
$i_2(t) = -0.558\sin 50t - 3.100\cos 50t + 4.598e^{-20t} - 1.498e^{-80t}$.

(b) $q_3(t) = 0.04(2 - 3e^{-50t})\sin 50t + 0.04(1 - e^{-50t})\cos 50t$;
$i_2(t) = -8(1 - e^{-50t})\cos 50t + 4(1 + e^{-50t})\sin 50t$.

(c) $q_3(t) = 0.1\sin 50t - 5te^{-50t}$; $i_2(t) = 5(1 + 50t)e^{-50t} - 5\cos 50t$

16.(i)

$$\begin{aligned} m(t) &= \frac{\alpha}{3}(1.103e^{-0.001t} - 0.436e^{-0.036t} + 0.333e^{-0.009t}); \\ f(t) &= \frac{\alpha}{3}(1.36e^{-0.001t} - 0.36e^{-0.036t}); \\ w(t) &= \frac{\alpha}{3}(0.8e^{-0.001t} + 0.2e^{-0.036t}). \end{aligned}$$

(a) $N(10) = 1.040\alpha$;
(b) $(m; f; w) = (35.8\%; 35.1\%; 29.9\%)$

16.(ii)

$$
\begin{aligned}
m(t) &= \frac{\alpha}{3}(1.123e^{0.018t} - 0.123e^{-0.075t}); \\
f(t) &= \frac{\alpha}{3}(1.123e^{0.018t} - 0.123e^{-0.075t}); \\
w(t) &= \frac{\alpha}{3}(0.935e^{0.018t} + 0.065e^{-0.075t}).
\end{aligned}
$$

(a) $N(10) = 1.241\alpha;$

(b) $(m; f; w) = (34.6\%; 34.6\%; 30.9\%)$

17. Let $i(t), r(t)$ and $h(t)$ denote the infectious, the recuperating, and the healthy people at time t, respectively. Assume that the duration of the epidemic is short enough so that the natural increase of the population does not play a role, with the result that the total number of people N remains constant. If we also assume that the recuperating people are immune during the duration of the epidemic, then

$$h'(t) = -ah(t)i(t), \qquad h(0) = N - i(0);$$

$$i'(t) = ah(t)i(t) - bi(t), \qquad i(0) = c;$$

$$r'(t) = bi(t), \qquad r(0) = 0;$$

where the constants a, b, and c are positive real numbers. If the second assumption is not valid, then an extra term $r(t-d)$ must be added to the first and subtracted from the third equation, where d denotes the fixed period of immunity. Laplace transforms cannot be used because of the product hi in the first and second equations.

19. At time $t = \frac{1}{2\delta}\ln(1 - \frac{2\beta\delta}{c\alpha})$ the population of Nerds will be zero.

Section 6.5

2. For $0 \leq t \leq b$: $x(t) = \frac{a}{k}[1 - e^{\frac{-ct}{2m}}(\cos\frac{t}{2} + \frac{c}{m}\sin\frac{t}{2})]$
For $t > b$: $x(t) = e^{\frac{-ct}{2m}}(A\cos\frac{t}{2} + B\sin\frac{t}{2})$
where $A = \frac{2a}{k}[e^{bc}2m - (\frac{1}{2}\cos\frac{b}{2} - \frac{c}{2m}\sin\frac{b}{2}) - \frac{1}{2}(\cos\frac{b}{2})^2 + \frac{c^2}{2m^2}(\sin\frac{b}{2})^2] - \frac{4a}{m}(\sin\frac{b}{2})^2$
and $B = \frac{4a}{m}\sin\frac{b}{2}\cos\frac{b}{2} + \frac{2a}{k}[e^{\frac{bc}{2m}} - \cos\frac{b}{2} - \frac{c}{m}\sin\frac{b}{2}][\frac{c}{2m}\cos\frac{b}{2} + \frac{1}{2}\sin\frac{b}{2}]$

4.(a) $10^{-5}(2e^{-500t} - e^{-1000t});$

(b) $10^{-5}(1 - 500t)e^{-500t};$

(c) $10^{-5}(\cos 200t + 0.5 \sin 200t)e^{-100t}$.

5. (i) $q(0.001) = 1.5468 \times 10^{-6}$ farad;
(ii) $q(0.01) = 2.9998 \times 10^{-6}$ farad; $q \to 3 \times 10^{-6}$ farad.

6. $0.11 \cos(120\pi t + \delta)$ with $\tan \delta = \frac{386}{360\pi}$.

9. A solution must satisfy $y = 4x$ for $t \geq 0$. If $y(0) = 4$, then there is the solution $x(t) = e^{\frac{2}{3}t}, y(t) = 4e^{\frac{2}{3}t}$

10. For the cubic $a_3 s^3 + a_2 s^2 + a_1 s + a_0$ with each coefficient positive: $a_1 a_2 > a_0 a_3$. (This is a special case of a more general result, called the Routh-Hurwitz criterion - see, for example, [5] page 214.)

11. $\frac{80}{69} \cos 5t$

Section 7.5

1.(a) $x(t) = 1 - t \arcsin t - \sqrt{1 - t^2}$;

(b) x'(and hence also x) is unique on [0,0.86] by Theorem 3.2.1;

(c) No; x'' does not exist at $t = 1$.

3.(i) 5.89 m/s

(ii) 7.93 m/s

5.(a) $$a_7 = -\frac{w^2 + 40,320 a_3 a_5}{5,040(c + 8a_1)};$$

$$a_9 = -\frac{w^2 - 3,628,800 a_3 a_7 - 1,814,400 a_5^2}{362,880(c + 10a_1)}$$

(c) No.

10

References

[1] AKASHI, H. and LEVY, S., *The motion of an electric bell.* American Math. Monthly **65**(1968), 255-259.

[2] ALLEN, R.G.D., *Mathematical Economics.* MacMillan, London, 1956.

[3] APOSTOL, T.M., *Calculus,* Vol. 1. Blaisdell Publishing Co., New York, 1961.

[4] AYALA, F.J., GILPIN, M.E. and EHRENFELD, J.G., *Competition between species: theoretical models and experimental tests.* Theor. Pop. Biol. **4**(1973), 331-356.

[5] BELL, E.T., *Men of Mathematics.* Penguin Books, London, 1953.

[6] BELLMAN, R., *Dynamic Programming.* Princeton University Press, Princeton, NJ, 1957.

[7] BELLMAN, R., *Modern Elementary Differential Equations.* Addison-Wesley, Reading, Mass., 1968.

[8] BENDER, E.A. and NEUWIRTH, L.P., *Traffic flow: Laplace transforms.* American Math. Monthly **80**(1973), 417-423.

[9] BIRKHOFF, G. and ROTA, G.-C., *Ordinary Differential Equations.* Ginn, Boston, Mass., 1962.

[10] BOYCE, W.E. and DIPRIMA, R.C., *Elementary Differential Equations and Boundary Value Problems.* 2nd Ed. John Wiley, New York, 1989.

[11] BRAUER, F. and SÁNCHEZ, D.A., *Constant rate population harvesting: equilibrium and stability.* Theor. Pop. Biol. **8**(1975), 12-30.

[12] BURGHES, D.N., *Population dynamics. An introduction to differential equations.* Int. J. Math. Educ. Sci. Technol. **6**(1975), 265-276.

[13] BURLEY, D.M., *Mathematical model of a kidney machine.* Mathematical Spectrum **8**(1975/76), 69-75.

[14] CARLSON, A.J., IVY, A.C., KRASNO, L.R. and ANDREWS, A.H., *The physiology of free fall through the air: delayed parachute jumps.* Quarterly Bulletin, Northwestern University Medical School **16**(1942).

[15] CHURCHILL, R.V., *Operational Mathematics.* 3rd Ed. McGraw-Hill, Kogakusha, Tokyo, 1972.

[16] COURANT, R. and JOHN, F., *Introduction to Calculus and Analysis.* Vol. 1. Interscience, New York, 1965.

[17] DAHLQUIST, G. and BJÖRCK, A., *Numerical Methods.* Prentice-Hall, Englewood Cliffs, NJ, 1974. (Translated by N. Anderson.)

[18] DAVIS, H.T., *Introduction to Nonlinear Differential and Integral Equations.* Atomic Energy Commission, U.S.A., 1960.

[19] DUNCAN, O.D. and DUNCAN, B., *The Negro Population of Chicago.* University of Chicago Press, Chicago, Ill, 1957.

[20] ENGEL, J.H., *A verification of Lanchester's law.* Operations Research **2**(1954), 163-171.

[21] GENIN, J. and MAYBEE, J.S., *Introduction to Applied Mathematics.* Vol. 1. Holt, Rinehart, & Winston, New York, 1970.

[22] GILPIN, M.E. and AYALA, F.J., *Global models of growth and competition.* Proc. Mat. Acad. Sci. U.S.A. **70**(1973), 3590-3593.

[23] GORDON, S.P., *A family of generalized logistic curves and inhibited population growth.* Int. J. Math. Educ. Sci. Technol. **22**(1991), 919 - 925.

[24] GREEN, S.L., *Advanced Level Applied Mathematics.* 5th Ed. University Tutorial Press, London, 1970.

[25] GRODZINS, M., *Metropolitan Segregation.* University of Chicago Press, Chicago Ill, 1957.

[26] GURTIN, M.E. and MACCAMY, R.C., *Non-linear age-dependent population dynamics.* Arch. for Rat. Mech. and Anal. **54**(1974), 281-300.

[27] GURTIN, M.E., *Some mathematical models for population dynamics that lead to segregation.* Quarterly of Applied Mathematics **32**(1974), 1-9.

[28] HABERMAN, R., *Mathematical Models.* Prentice-Hall, Englewood Cliffs, NJ 1977.

[29] HANDELMAN, G.H., *A simple model of stability of structures.* SIAM Review **17**(1975), 593-604.

[30] HAUSRATH, A.R., *Stability properties of a class of differential equations modelling predator-prey relationships.* Math. Biosciences **26**(1975), 267-281.

[31] HAYES, P., *Mathematical Methods in the Social and Managerial Sciences.* John Wiley, New York, 1975.

[32] HENRICI, P., *Discrete Variable Methods in Ordinary Differential Equations.* John Wiley, New York, 1962.

[33] HUETER, G.J., McCLELLAND, M.A., RESNER, L.A., and ZEVALLOS, M.G., *An application of the Lanchester model to the Battle of the Ardennes.* Interface **5**(1978), 15 - 25.

[34] IDYLL, C.P., *The anchovy crisis.* Scientific American **228**(1973), 22-29.

[35] KAPLAN, W., *Operational Methods for Linear Systems.* Addison-Wesley, Reading, Mass. 1962.

[36] LAMBE, C.G. and TRANTER, D.J., *Differential Equations for Engineers and Scientists.* The English University Press, London, 1961.

[37] LAMBE, C.G., *Applied Mathematics for Engineers and Scientists.* The English University Press, London, 1958.

[38] LANCHESTER, F.W., *Aircraft In Warfare: The Dawn Of The Fourth Arm.* Constable and Co. Ltd., London, 1916.

[39] LANCHESTER, F.W., *Mathematics in Warfare.* The World of Mathematics, Vol IV, 2138-2157. Simon and Schuster Inc., New York, 1956.

[40] LOTKA, A.J., *Elements of Physical Biology.* Williams & Williams, Baltimore MD, 1924. (Reprinted as Elements of Mathematical Biology. Dover, 1956.).

[41] MAKI, D.P. and THOMPSON, M., *Mathematical Models and Applications.* Prentice-Hall, Englewood Cliffs, NJ, 1973.

[42] MATHEWS, J.C. and LANGENHOP, C.E., *Discrete and Continuous Methods in Applied Mathematics.* John Wiley, New York, 1966.

[43] McGILL, D.J. and KING, W.W., *An Introduction to Dynamics.* PWS-Kent Publishing Co., Boston, Mass, 1989.

[44] MILLER, R. and BOTKIN, D., *Endangered species: models and predictions*. Amer. Sci. **62**(1974), 172-181.

[45] MURRAY, J.D., *Some simple mathematical models in ecology*. Mathematical Spectrum **16**(1983), 48-54.

[46] NOBLE, B., *Applications of Undergraduate Mathematics in Engineering*. The Mathematical Association of America. MacMillan, New York, 1987.

[47] PECSOK, R.L. and SHIELDS, L.D., *Modern Methods of Chemical Analysis*. John Wiley, New York, 1968.

[48] PELLA, J.J. and TOMLINSON, P.K., *A generalized stock production model*. Bull. Inter-Am. Trop. Tuna Comm. **13** (3)(1969), 421-458.

[49] PIELOU, E.C., *An Introduction to Mathematical Ecology*. Wiley-Interscience, New York, 1969.

[50] RITGER, P.D. and ROSE, NJ, *Differential Equations with Applications*. McGraw-Hill, New York, 1968.

[51] ROBERTS, C.E., Jr., *Why teach existence and uniqueness theorems in the first course in ordinary differential equations?* Int. J. Math. Educ. Sci. Technol. **7**(1976), 41-44.

[52] ROWE, R.R., *The vicious circle*. Journal of Recreational Mathematics **5**(1972), 207-210.

[53] SALTZ, D., *An Introduction to Analysis*. Prentice-Hall, Englewood Cliffs, NJ, 1965.

[54] SAUER, N., *Private communication*.

[55] SMITH, F.E., *Population dynamics in Daphnia*. Ecology **44**(1963), 651-663.

[56] SOKOLNIKOFF, I.S. and REDHEFFER, R.M., *Mathematics of Physics and Modern Engineering*. McGraw-Hill, New York, 1958.

[57] SOKOLNIKOFF, I.S. and SOKOLNIKOFF, E.S., *Higher Mathematics for Engineers and Physicists*. 2nd Ed. McGraw-Hill, New York, 1941.

[58] SPIEGEL, M.R., *Applied Differential Equations*. 2nd Ed. Prentice-Hall, Englewood Cliffs, NJ, 1967.

[59] STOKER, J.J., *Nonlinear Vibrations in Mechanical and Electrical Systems.* Interscience, New York, 1950.

[60] STROGATZ, S.H., *Love affairs and differential equations.* Mathematics Magazine **61**(1988), 35.

[61] TEICHDREW, D., *An Introduction to Management Science.* John Wiley, New York, 1964.

[62] TENENBAUM, M. and POLLARD, H., *Ordinary Differential Equations.* Harper and Row, New York, 1963.

[63] VAN DER PLANK, J.E., *Dynamics of epidemics and plant disease.* Science **147**(1965), 120-124.

[64] WIDDER, D.V., *Advanced Calculus.* 2nd Ed. Prentice-Hall, Englewood Cliffs, NJ 1961.

Index

— NOTES —

— NOTES —

— NOTES —